14,95

ESKIMO YEAR

Eskimo Year

BY GEORGE MIKSCH SUTTON

Illustrated with Drawings and Photographs by the Author

With a Foreword by David F. Parmelee

UNIVERSITY OF OKLAHOMA PRESS
NORMAN

OTHER BOOKS BY GEORGE MIKSCH SUTTON:

Bird Student: An Autobiography (Austin, 1980)
To A Young Bird Artist: Letters from Louis Agassiz Fuertes to George Miksch Sutton (Norman, 1979)
Fifty Common Birds of Oklahoma and the Southern Great Plains (Norman, 1977)
Portraits of Mexican Birds: Fifty Selected Paintings (Norman, 1975)
At a Bend in a Mexican River (New York, 1972)
High Arctic (New York, 1971)
Oklahoma Birds: Their Ecology and Distribution, with Comments on the Avifauna of the Southern Great Plains (Norman, 1967)
Iceland Summer: Adventures of a Bird Painter (Norman, 1961)
Mexican Birds: First Impressions (Norman, 1951)
Birds in the Wilderness (New York, 1936)
The Exploration of Southampton Island, Hudson Bay (Pittsburgh, 1932)
An Introduction to the Birds of Pennsylvania (Harrisburg, 1928)

The paper in this book meets the guidelines for permanence and durability of the Committee on Production Guidelines for Book Longevity of the Council on Library Resources, Inc.

Library of Congress Cataloging in Publication Data

Sutton, George Miksch, 1898–1982
    Eskimo year.

    1. Eskimos—Northwest Territories—Southampton Island. 2. Natural history—Northwest Territories—Southampton Island. 3. Southampton Island (N.W.T.)—Description and travel. 4. Sutton, George Miksch, 1898–1982. I. Title.
E99.E7S88   1985   971.9'2   84-28086
ISBN 0-8061-1933-0 (alk. paper)

Dedication

WHEN Jack Ford reads this page he will laugh like a wild loon. He will be frightfully embarrassed. He will wonder why any book, after all, should be dedicated to anyone—especially a book of this sort. But he will turn to this page now and then in spite of himself. And a smile will battle for first place in his eyes. For he will remember the happy-faced Eskimos, the Huskies and their howling, the *netchek*-seals, the caribou, the white foxes, the gulls, the moss-flowers, the sculpins—all the beautiful beings that live on the island that is called Shugliak; and he will remember "the Doctor" who became his friend during the winter of 1929–1930.

Foreword to the Second Edition

THE AIVILIKMIUT ESKIMOS were well aware of our western customs when George Miksch Sutton arrived at Shugliak. Even so, apart from the Hudson Bay Posts, missions, and annual visits by big ocean freighters, their day-to-day existence continued to be really primitive. Living and traveling with these hardy, walrus-hunting Eskimos, Sutton was catapulted into an unyielding, mystic world. From this experience evolved *Eskimo Year*—an unusual blending of a unique people with a unique wildlife. Although Sutton claims it to be only the record of a glorious adventure and not an ethnological treatise, his narrative is bound to attract the anthropologist, biologist, and historian as well as the adventurer. Having read the book as a youngster, I was immediately and forever imbued with a polar obsession.

It was my good fortune in 1953 to be introduced to the Canadian Arctic Islands by none other than George Sutton. In quest of Greenland wheatears and other tundra birds at the very edge of retreating snows, we crisscrossed the undulating slopes of Baffin Island's Frobisher Bay. Nesting snowy owls were abundant that season, thriving on a super-abundance of lemming; in later years Sutton often reminisced on the fierce attacks by the parent owls in defense of their young. When winter's ice finally left Frobisher in midsummer, we sailed with the Eskimos in Hudson Bay boats to the innermost extremities of the bay. An unforgettable incident on one of those trips was being lost at sea in dense fog. After we had reached a point of near desperation, the distant wailings of Eskimo dogs led us back through the icy mist.

To this day I can mentally hear those wailings—as only tethered huskies can wail.

Our next trip with the Canadian Eskimos began nine years later on June 2. We had been in the Cambridge Bay area of Victoria Island since May 10, and having experienced almost no thaw, were overjoyed to see our first sunny weather on the eve of departure. Our destination was a string of offshore islets called the Finlaysons, located about thirty sledge miles west of Cambridge Bay and not far from a landmark known as Starvation Cove. The plan was to proceed overland to the Augustus Hills, where, we had been told, there was an abundance of birds. From there we would sledge southwestward over land and sea ice directly to the Finlaysons, thence by sea ice all the way back to Cambridge Bay. Since there were six of us, two sledges were necessary.

The Eskimo guides were eager to go by early afternoon. Our two companions, Steve Stephens and Richard Schmidt, piled onto the sledge of a famous Cambridge Bay hunter, David Koomyuk, while Sutton and I scrambled onto the sledge of another, called Peter Panaktuk. The long whips cracked, the dogs howled, and the sledges seemingly flew over hard snow broken occasionally by a withered willow top. Progress was swift until we reached the Hills which lay largely exposed by a rapidly advancing thaw. The sledges slowed on a sluggish trail of mushy snow.

At once we were met by a hailstorm of migrating birds. Koomyuk was right, this was a birdy place! On the ground were more rock ptarmigans than Sutton and I had ever seen. Overhead were flying flocks of golden and black-bellied plovers, swans, eiders, and geese of several species. A myriad of larks, buntings, and longspurs were busy performing pretty flight-songs. In the early evening hours the sledges ground to a halt, and up went the tents in the only dry spot within a quagmire of melt pools. Soon all of us fanned out over the low-lying hills, whose peculiar flat tops were like miniature mesas.

In the late evening hours, at a time when the sun was low on the northern horizon and the tundra colors enchanting, I caught up with Sutton, who stood transfixed, looking far out over the denuded, rosy-tinted hills and icy sea beyond. Without turning his head he said this had to be the wildest, loveliest tundra scene he had ever seen. Later, I realized that at that very moment he envisioned a painting of a stately flight of Arctic cranes that only moments before had descended several ancient beach terraces.

Steve Stephens returned to camp excited about having seen a golden eagle chased hard by a falcon of uncertain identity. I had seen nothing that spectacular, but my thoughts and spirits were so full of every kind of bird that I failed to sleep a wink. None of us slept very well, because our fascination for those hills was almost unbearable. Several miles from our tents, Sutton and I came upon an Eskimo encampment that was nearly ankle-deep in bird remains. Scores of ptarmigan, many still in handsome winter feather, were being shot and, to our horror, tossed to the dogs. Within the Hudson Bay tents the Eskimos slept on caribou skins placed on beds of snow. Sutton mused that not all native customs had disappeared.

By the following noon it was time to leave the hills for the Finlaysons, where, two years before, Schmidt and I had come upon nesting Thayer's and glaucous gulls. Soon we were gliding over a frozen sea that was nearly free of troublesome rubble ice and snow. Inevitably these sledging trips at sea were interrupted for frustratingly longsome periods. Whenever a seal was found snoozing on the ice beside its breathing hole, our Eskimos could not resist going after it despite our vehement objections. And so it was this day when the two sledge teams separated, suspiciously by design. Our guide, Panaktuk, had spotted a seal a long way off, a mere dot on the horizon. With his belly flat on the ice and rifle held across his arms, he set off on a long, tortuous, seemingly endless crawl that stopped abruptly every time the seal raised its head. The dogs, of course, sensed a kill and howled incessantly. They could not run behind the hunter because before leaving us Panaktuk had struck the heavy sledge anchor deep in the ice. The dogs yelped, bit one another, pulled and strained at the traces. The sledge vibrated and lurched nearly out of control.

Try as we did, Sutton and I failed to calm the dogs. By the time Panaktuk approached his quarry, the ice anchor let go. I glanced at Sutton who said nothing, for his eyes said it all. How we survived the mad race to the seal hole, I'll never know. We zigzagged back and forth out of control, hanging on for dear life and miraculously missing gaping cracks in a deteriorating ice. Naturally, the dogs frightened off the seal before Panaktuk had made his kill. Fortunately for us, the dogs ran for the abandoned hole, where Panaktuk had no trouble catching the runaways—and exercising his great displeasure with them.

The Finlaysons loomed up stark and naked, black and forbidding in a lonely sea seldom visited by sledge or boat. The vociferous gulls were there as expected, but the snow on their nesting ledges was so thick that none had started to nest. An old raven's nest close by appeared unoccupied, but not so a peregrine's eyrie that commanded Sutton's studied attention. With help from the Royal Canadian Mounted Police, Sutton and Stephens returned by boat later that year and captured five baby Thayer's gulls, which they raised by hand. The smallest one they dubbed "Pevee." The master artist painted all five directly from life and inscribed on this treasure "Pevee and Friends...Drawn for Helen Gale Parmelee at Cambridge Bay, Victoria Island, September, 1962, by her friend George Miksch Sutton." Helen Gale was seven years old.

Our trip back to Cambridge Bay was swift, monotonous, without incident. It lasted throughout the twilight night. All of us were exhausted upon arriving at base camp, not so much physically as from the excitement of it all. Sutton didn't know then, nor did any of us, that the Finlayson trip would be our last sledging. Snowmobiles soon replaced the time-honored sledges. The trip also was Sutton's last with the Eskimos. Throughout George Sutton's many productive years, the north country remained up front in his thoughts. He frequently quoted his Eskimo friends and not once did he lose his youthful admiration for them. This love all comes out in *Eskimo Year.*

DAVID F. PARMELEE
Professor and Program Director
Field Biology Program
University of Minnesota

Contents

Illustrations

Muckik pointing out caribou country to John Ell.

A sledge loaded with caribou.

Arctic fox caught in a trap.

"Viscount Grey" a pet Collared Lemming.

Weasel.

Little Peter, whose Aivilik given name is obscene.

John Ell and Muckik.

Aivilik girls at the entrance of their igloo.

Between pages 246–47

Aivilik hunter with his harpoon.

By and By and Tommy Bluce with Eskimo children.

Doctor Sutton, far right, shown with John Ford and a walrus aboard ship.

Home from Southland: an early Snow Bunting.

Herring Gulls.

Eider nest.

Willow Ptarmigan cock and hen.

Red-throated Loon on nest.

Guillemots.

A chronicle of changing seasons in the mother Willow Ptarmigan's back.

Willow Ptarmigan chick.

A nest of young Snowy Owls with a lemming nearby for food.

Dryas blooming in the wake of the retreating snow.

Aivilik village of igloos and tupeks in the early spring.

An early spring nest with six Snowy Owl eggs.

The photographs listed above were reproduced from Dr. Sutton's tinted, glass-mounted slides of scenes on Southampton Island. Dr. William R. Johnson prepared the legends.

ESKIMO YEAR

Map of North America showing Southampton Islands location in Hudson Bay.

CHAPTER I

Shugliak: the Island that is called Southampton on the Charts

THERE is a nineteen-thousand-square-mile heap of rock that sprawls almost across the mouth of Hudson Bay. The Eskimos that inhabit this vast heap of rock call it Shugliak: The Island-Pup That Is Suckling the Continent Mother-Dog. They never call it Southampton Island, for to them Southampton Island is simply not its name. They cannot read the proud words on the White Man's charts; and when they hear the White Man talking about Southampton it does not occur to them that Shugliak and Southampton can be one and the same place.

I am one of the few white men that have lived on Shugliak. The Island-Pup That Is Suckling the Continent Mother-Dog was my home for a year. I am obliged to call the place Southampton Island rather than Shugliak most of the time because I wish not to be misunderstood by cartographers, zoögeographers, and casual acquaintances. But you will understand me when I say that it was Shugliak, rather than Southampton Island, that was my home; that it is to Shugliak, not to Southampton Island, that I wish to return. I am not sure, myself, that Southampton Island and Shugliak are precisely one and the same. Shugliak is a place where interesting and friendly human beings live; Southampton Island is part of a map.

Shugliak has a desolate appearance, but it is not desolate.

You go there to gather data upon Arctic mammals and birds and insects and plants, and in the midst of your collecting of specimens you find yourself absorbed in observing the brown-skinned folk about you, who call themselves the Innuit.[1] You expect to be lonely and are half disappointed at finding yourself lonely so little of the time. You expect to write philosophical discourses during the long, "shut-in" winter and you find yourself playing boisterous games. A strange island, this Shugliak: strange, but not desolate.

Shugliak is very cold in winter, so cold the ground cracks with a loud sound and the open water of the ocean steams. But in summer there are white and yellow and purple flowers all over the meadows and ridges; and butterflies and bumble-bees and crane-flies; and uncounted and not to be counted mosquitoes that rise like smoke from the marshes during the short warm season that is called *Mittiadlut*.

The only trees are stunted willows and tiny birches that are buried under the snow every winter. But in summer the leaves of these lowly trees cast a delicate shadow-lace on the rocks and moss that you can see if you get down on your hands and knees, and they turn bright orange and wine-color with the first sharp frosts of late August.

Back in the days of Button and Baffin and Hudson no one but the Eskimos knew that Shugliak was an island. White men called the all but unknown region to the northwest of Hudson Bay Southampton Land. Not until 1742, and the voyage of Captain Christopher Middleton, and the discovery of Frozen Strait just south of Melville Peninsula was Shugliak's insularity definitely established. It was at this time that the name Southampton Island came into use among the so-called civilized peoples of the South Country.

Shugliak is a grand old island, flat and yellow-brown and

[1] Eskimos.

nothing but chunks of limestone all over the western half; and high and rough, with granite cliffs, over most of the eastern half. Herds of caribou roam the wilderness barrens; polar bears abound; walrus and seals thrive in the icy waters. Among the innumerable grass-lined lakes water-birds of many species nest.

Two races of Eskimos live on Shugliak today. They call themselves Aivilikmiut and Okomiut. They are not, strictly speaking, native to the island, for they were brought from Repulse Bay and from Baffin Island by the early whalers and by the Hudson's Bay Company. The indigenous race were known as the Shugliamiut,[1] the Innuit living on Shugliak. This fine people died out in 1902, victims of a savage and mysterious epidemic.

The Eskimos are childlike, jolly, lovable. They shake hands with you and try to talk to you and bring you walrus-ivory carvings and mend your clothes and make sealskin boots for you and teach you how to drive the dog-team and how to crack the walrus-hide whip and how to hunt caribou. They know ever so much more than you do about life and living and religion and philosophy. They know when to think and when to not-think. They love one another in a helpful, practical, tender, but not sibby-sobby, sentimental way. Their language sounds primitive but it probably isn't primitive. They don't bathe much, if any, but they aren't really dirty. They have an odd odor, but it is a happy odor. When you come to know the Eskimos you wish you were an Eskimo, not just an outsider looking on. You say to yourself: "But I belong to a superior race!" and then you realize that you only think you are superior, that you call yourself superior only to help keep

[1] By ethnologists and anthropologists this word is frequently spelled *Saglernmiut*. As pronounced by the Aivilik Eskimos, however, it is Shugliamiut.

up your courage, that you really wish to God you didn't have
to keep on thinking superiority so hard all the time . . .

There are about one hundred and fifty Eskimos on Shugliak today: men, women and children all counted. They just
about hold their own from year to year: an old man dies; a
baby is born.

The white population is practically inconsiderable. While I
was on the island, five whites, including myself, lived there:
two traders, two French missionaries, and a scientist—all
men. Today there are two traders (one with a wife and two
children), two missionaries, and a young English geographer.

There are two missions at Shugliak's little trading post at
Coral Inlet: the Roman Catholic and the Anglican. The Eskimos attend the mission services sporadically, enjoying the
candles, the ceremonies, the pictures, and the singing. Sometimes they are "converted." But an Eskimo has a loyalty to
established folk-ways that fortunately does not change with
superficial gestures. He may "become" a Christian, but he is
first of all, and unalterably, an Eskimo. Strange, this need for
converting to Christianity of a people whose daily behavior
in *tupek* and *igloo* was biologically so sound before the White
Man came. Strange, this teaching Christ to persons who are
intrinsically so gentle, so happy, so childlike. Strange, this
hope that the Innuit may accept a religion that has not developed on the tundra, that knows so little of *igloo* building,
of walrus hunting, of dog-team driving. Strange, this urge to
educate and improve and change a people that have evolved
for themselves a dignified culture, an adequate religion, and,
above all, a serene contentment.

Shugliak is the home of the Arctic fox. These foxes the
Eskimos trap all winter, bringing the snowy pelts in for trade
from time to time. Polar bear-skins, baleen or whalebone, and
white porpoise skin also are traded and exported. Walrus

hide used to be an important item, but the Canadian Government wisely has stopped the killing of walrus save for dogfood.

Winter is the time you ought to be lonely on Shugliak. Occasionally you are or think you are. But most of the time you are not. Your friends won't let you get that way. There is no end of hunting and trapping and travelling, when the weather is fine. When blizzards shut you in there are games and wrestling matches, salted specimens to be worked over, drawing- and writing-work to be done. At the post there is candy-making or pie-baking, and the radio, and the organ at the Mission. And there are those ten books you somehow hope to read: the dictionary, Shakespeare, Leaves of Grass, Fathers and Sons, The Romance of Leonardo da Vinci, The Oxford Book of English Verse, The Cruise of the Neptune, The Birds of Western Canada, *Pêcheur d'Islande,* and Stevenson's glorious *Aes Triplex.*

Summer is good fun. But it is always so bright and there is so much to do you go half crazy trying to decide what to do first and whether it's today or tomorrow or still only yesterday.

Chapter II

South Bay and Coral Inlet

On the afternoon of Friday, August 16, 1929, the Hudson's Bay Company's supply-ship *Nascopie* steamed past the flat western end of Coats Island, then close by a rough, high rock named Walrus Island, at the entrance to South Bay.

What? You've never heard of Walrus Island? Of course you've heard of Walrus Island. The North Country's full of rocks named that. Every strait, every gulf, every bay, every inlet, every sound in the Arctic has its Walrus Island. Some of these are called Walrus Island because walrus are to be found thereabouts; others are called Walrus Island because walrus never are to be found thereabouts; and others are called Walrus Island, I suppose, after a man named Walrus, or a man who looked like a walrus, or a man who liked raw clams. It is no easier accounting for names in the North Country than it is anywhere else.

The Walrus Island I am telling you about is a special one, of course. For it is near old Shugliak. And there really are walrus there. That Friday evening, as I leaned on the railing of the *Nascopie* looking at Walrus Island, my elbows tingled. High on a ledge, head all golden in the slant sunlight, lay a polar bear. How majestic his mien! What superb indifference in his easy pose! Already we were in Fisher Strait, already in South Bay. Walrus Island loomed now, above our wake. Southampton of the charts was off there in the grayness to the north—somewhere just beyond the horizon's rim. Southampton of the charts and of unnumbered dreams: the island that was to become old Shugliak on the morrow.

I went to sleep that night, but going to sleep was not easy. There was too much to think about. We had had such a glorious trip: thirty happy, friendly days—on the Saint Lawrence, down the Labrador, across Ungava Bay, in Hudson Strait, at last in Hudson Bay. Captain Murray was such a bully good captain. "Chiefy" Leddingham was so grouchy but so swell. "Doc" Clothier was such a jolly pal. That guy Hardwick— what a man! What would this place called Southampton be like anyway? Would Sam Ford remember me, or would my coming get him all upset and rattled? Would he and I end up the year friends or enemies? Had my equipment come through all right? How was everybody back home by now?

That party at Port Harrison. What a grand lot of ruffians these North Country boys were anyway! Jimmy Thom! Not so easy saying good-bye to a chap like Jimmy, realizing you'll probably never see him again, that you'll probably never even get a letter from him. What a violent animal he was, anyway, liking you so thoroughly, despising you so utterly, using his fists and saving words. But a chummy person too, who liked to talk things over. Who was it, at that Harrison party, that took his false teeth out and dressed them up in a bit of silk and passed them round? Who was it that persisted in drinking his beer through the loud-speaker? Who was it that whimpered over being hit on the head with a tablespoon and for sympathy got biffed in precisely the same spot with a bottle? What was that song we all tried to sing? Was it Barnacle Bill with the worst words we could think of, or was it Carry Me Back to Ol' Virginny, or was it Sonny Boy? "Doc" Clothier could sing Sonny Boy awfully well . . .

When I looked out of the port-hole next morning, there,

beyond the green water that was Coral Inlet, lay Southampton, a broad, yellow-brown band hung with fog. There were the little white buildings of the trading post, clustered on a low, rocky point. A long-tailed jaeger was circling the *Nascopie,* half-heartedly chasing the terns and waiting for garbage to be thrown from the galley. Everywhere was a *putt-putting* of motor-boats, a rattling of windlasses, a jargon of voices shouting out commands.

Sam Ford, Chief Trader at the post, came in to greet me before I had got out of my bunk. "Well, Doc, I wondered if you'd come," he said, and his pale gray eyes shone. So Sam had remembered me. It was good to see Sam again. Then came Jack, nineteen-year-old Jack, Sam Ford's son, a lad who was to become my friend and all but constant companion. When Jack began telling me about the swans and geese and cranes he had seen on the island we forgot everything else; we sat there like a couple of fellows who had gone to college together and hadn't seen each other for years. We talked as if we had only an hour for saying our say: as if there couldn't possibly be twelve quiet months together ahead of us! Finally I jerked on my clothes and came on deck.

NOWYAH: HERRING GULLS

I shook hands with Eskimos and Eskimos and Eskimos. I liked them all immediately (you can't help liking an Eskimo!), but felt a little shy with them. I wanted my hunting outfit. I wanted my gun. This gray suit was too neat. I mean it really was too neat for this country, too neat and trim.

Some sort of feast was going on. A great dishpan of meat and biscuit had come from the galley and the Eskimos were diving in. What a strange lot they were! Where had they got these caps, these hats, these creaking tan shoes, these gay silk handkerchiefs, these pale blue suspenders? What right had an Eskimo to curly hair, or finely chiselled nose, or thin lips, or green-brown eyes? I shook hands with them over and over again, grinning foolishly and wishing I could tell them how happy I was. Somebody had told me how to greet an Eskimo —the proper words to use, that is—but I couldn't bring myself to saying anything until I knew exactly what the words meant. They must have thought me a very peculiar white man. But they knew who I was. They knew I had come to their Shugliak to look for birds. The talk had gone round, you see.

We went ashore. There was more shaking of hands. Everywhere there was talking and laughing and running. Some of the men were busy loading bales of fox skins and rolls of bear-skins and bundles of porpoise hide onto the *Nascopie*. Others were groaning and grunting under my ponderous chests of equipment that had been landed. Some were smoking pipes or cigarettes, or chewing gum.

Two bearded white men in black hats and long black robes walked up to greet me with friendly smiles. They were Father Thibert and Father Fafard, the Roman Catholic missionaries.

Eskimo women stood about in giggly groups, talking slushily between their cheeks and molar teeth. Funny sound, that talk of theirs! They had on tin-spangled, braid-decorated, blue velveteen dresses that swept and swished over the ground, duffel dickey-coats with flaps down the front and down the back, plaid shawls, sealskin boots, and rumpled red and black hair ribbons. One had got herself a pair of high-heeled shoes. She hobbled about among the rocks like a kitten in walnut-shells.

Children were running or standing about everywhere, most of them with solemn expressions on their faces. All this hubbub was too much for them. There were too many tall, fierce-eyed ogres; too many clean clothes; too many official records in everybody's hands. Some of the little girls had on long red dresses or funny, lumpy aprons. Their hair was usually braided tight. The little boys wore all sorts of clothing, some of them stiff trousers of moleskin open from front to back and held up with brightly colored braces. The bigger boys wore baggy trousers that stuck into their sealskin boots, orange- and brown-checked sweaters, woollen tam o'shanters, and bright lumberman's shirts.

I could hear a gramophone squeaking in one of the tents; the humming of a sewing machine; an accordion somewhere off in the distance; a thin-voiced harmonica.

The tents pitched there at the post were nearly all of canvas. One or two were of sealskin. These tents the Eskimos called their *tupek*.

Then I began hearing about the bear. "Be sure to give him plenty of air and some salt-water now and then. And a bath whenever you can. Fish and seal-meat is what he likes to eat." Where was this much-talked-about creature?

He was in a heavy crate at the water's edge, waiting to be put on the *Nascopie;* waiting to begin his long journey to a London zoölogical garden. He was as foul a creature as I ever saw, his coat so badly matted with grease and filth that he was the color of oil-barrels. He was not a pleasant brute either.

He looked as if he wanted to chew somebody up, to rake somebody's legs with his sledge-hammer forepaws, to disembowel some dog with an offhand slash or two. He padded back and forth ceaselessly, looking out toward the sea with his glittering, baleful, piggish eyes. The Eskimo children stood round at a respectful distance. The dogs, who longed to be at him, were stoned and clubbed whenever they came near.

Those dogs. You are wondering already why I should say "who" instead of "which" when I speak of the Huskies. Maybe you prefer "the dogs which longed to be at him" to "the dogs who longed to be at him." But I don't. Dogs are almost human in the North Country. They are too much a part of going anywhere, too much a part of hunting anything, too much a part of life and living to be classed with lemmings, pebbles and willow-twigs. You can say "a lemming which" and "a pebble which" and "a willow-twig which"; but you can't say "a dog which." Not when the dogs love you as they do, and when you love the dogs as you do. Not love in any well-known sense of the word, of course. Not the love you have for a fox terrier or an Airedale. You love dogs in the North Country somewhat as you love life, in an impersonal way; and it is thus that the dogs love you. You and the dogs depend on one another. It is this symbiosis that makes existence possible. Huskies are the meanest brutes ever, upon occasion. You hate them at times; but you love them more. You beat them frightfully, but you do not kill them. Any more than you kill yourself. For killing your dogs is killing yourself in the Arctic. Live on old Shugliak for one winter, and you will see.

Sam Ford showed me my workroom, upstairs in the little house where he and Jack lived. He showed me too the egg of a rare gull, the skin of a blue goose (a bird whose nesting-ground was at that time unknown to the Outside World), and some Arctic wolf hides. The Eskimos helped me unpack my cameras and guns and scientific materials. Doctor Clothier and I took a walk out over the tundra. We saw a spider, a bumblebee, a few flowers, some glossy red mushrooms, and a pair of Pacific loons. We got our feet wet in the spongy moss. On the way back we talked about our trip on the *Nascopie* and our separate plans for the winter. In his stateroom he had some sort of medicine the taste of which I liked very much. It was something on the order of cascara sagrada. He gave me a huge, fifteen-inch-high bottle of it, just as a friendly gesture. Sam and Jack and I drank a tablespoonful of this nice, brown syrup nearly every Saturday night during the winter, by way of going on a spree, or to the theatre, or to the club.

By noon of the next day I was beginning to recognize some of the Eskimos, and to remember one or two names. Old Shoo Fly somehow was the one I remembered most easily, her name was so funny and her fine old face so friendly and full of character. Then there was Shoo Fly's husband, whose Eskimo name was Angoti Marik, but who frequently was called Scotch Tom. And there was Amaulik Audlanat, the chief servant at the post, a strong, capable man usually called John or John Ell. And there were Cabin Boy; and By-and-By; and Mrs. One-eyed Joe; and Tommy Bruce or Tommy Bluce or Tommy Loose (I never did know which!); and Curly Joe; and green-eyed Pumyook; and Kooshooak; and young Santiana; and little Ookpik, John Ell's daughter; and Mary Ell, John's wife; and curly-headed Shookalook; and that indescribably wild-looking Khagak; and Sheeloo; and Tapatai; and the old widow Susie; and the middle-aged widow Kuklik; and so on and so on. The names were a jumble in my mind. I couldn't have written them down this way on that day, not if I had tried for hours; but they were all there, for it was ship-time, and I shook hands with them over and over again.

I told Sam I was surprised at seeing thin-lipped, curly-haired Eskimos on his island. Sam told me something of the Aiviliks and the Okomiut. He said the Aiviliks were a Re-pulse [1] Bay tribe and that some of them were not full-bloods. Tousled-headed Shookalook, for example, was the son of a Portuguese seaman. Pumyook's wife, an unusually pretty girl, was the daughter of an American whaler. Pumyook himself was partly white. And the Okomiut were the "Baffin Land-ers." They too were a hybrid race to some extent. But the Aivilikmiut and the Okomiut did not frequently intermarry.

I took note of the trading post. There were ten buildings, all of wood or principally of wood, and all painted white with red roof, save two. There was the big Hudson's Bay Company Store with its broad, familiar sign, the Chief Trader's house, the servants' house, two store-houses, and an oilshed; the Roman Catholic Mission building with its steeple and lofty cross on which the hawks liked to perch; the smaller Anglican Mission building, at that time without a cupola; the little powder magazine, way off at one side and painted red; and the strange pile of stones covered with boards and tin and iron that was called the Widow Kuklik's shack.

On Sunday, August 18, the *Nascopie* pulled out. There was something solemn about the way she became smaller and smaller and lower and lower and about the way her smoke showed so long after she herself had disappeared; but I was glad she was gone. The rockets that roared were not to me a sad farewell. They were a summons to work.

That evening Sam and Jack and I threw six boxes of candy (the hard-tack sort, with lemon drops and little striped cylin-

[1] Sam and Jack always pronounced this word Ree-pulse, not Re-pulse'; so *Ree-pulse Bay*, please.

ders and pillow-shaped pieces) to the Eskimos. We just tossed handfuls of the stuff into the air letting it fall where it would, in true North Country style. What a scrambling! What a screaming! What a helter-skelter running this way and that! And no stomach-aches either, so far as I could learn.

CHAPTER III

The Post

THOSE first days on old Shugliak weren't easy days. Not that I was homesick. I was anything but homesick. I did not in the least dread the thought of the year of isolation ahead of me. But I was so frightfully in earnest about the task I had set out to accomplish, so mad to get to work, so eager to prove to myself that this idea of a "solo expedition" to an Arctic island was not a foolish idea.

Shugliak was a big island. I had twelve months at my disposal for making a biological survey of these nineteen thousand square miles. All sorts of natural history specimens must be collected and preserved. Photographs must be taken; maps made; full records kept.

The bigness of the job almost overwhelmed me at first. How could I memorize the unpronounceable, unspellable names of these Eskimos who were to be my friends and companions? How could I make myself remember the names of their dogs? How could I ever learn to talk to them in their own tongue? I wanted to travel about with them, you see; I wanted to learn from them their ideas about the island; and what is more, I wanted to like them so well that I should enjoy being with them during the winter that was coming.

I got my workroom in order. John Ell built a whole wallful of shelves for me out of drygoods boxes. I felt better when my guns and cameras and water-color outfit were unpacked and ready for use, and when I got a list of Eskimo names written down in black and white so that I could see them.

Sleeping was a farce those first two nights. The Eskimos were whooping it up at all hours, endlessly playing games, screaming and laughing. I'd nap for a while, then waken with a start in the midst of some dream of packing equipment or running to catch a train. During sleepless hours I would memorize Eskimo names and say over Eskimo expressions such as those for "Good night," "Good morning," and "The wind is blowing hard today." You see I had had too easy a time on the *Nascopie;* those thirty days of idleness were telling on me now.

I determined not to work too hard. Not that I was afraid of falling ill from long hours; but I was not going to be queer. If this island was to be my home for a year, I must set out to contribute something of friendliness, something of gaiety, to every day.

We had some canned stuff for meals, just about what you'd expect, and bread that Mary Ell baked over at the servants' house. But bread and canned stuff weren't all we had. We had fried salmon trout, a fish the Eskimos called *ichalook,* and just about the finest fish I ever tasted. I never once tired of it, though we had it every meal for long stretches through the winter. And I swear I'm hungry for it right this minute.

Sam and Jack went into long huddles with the Eskimos about me. They explained in great detail just what I was and wasn't. I was a bird-man. I was just as much interested in little birds as in big birds. I needed birds, too; not birds to eat, but birds I could skin and take back "stuffed" to the White Man's Land. I was not a fur trader. I was interested in foxes, but I didn't want all the fox skins I could get. I was just as much interested in lemmings and weasels and hares as I was in foxes. I was a strange being.

Specimens soon began to arrive: half-grown phalaropes and plovers and sandpipers and baby old-squaw ducks with their heads bashed in and brains oozing out; a loon with a two-inch section of its neck shot away, the head neatly sewn back with caribou sinew, but sewn back upside down; lemming and weasel skins stuffed with grass; a crane skin, cleaned so thoroughly that all the important leg-bones were missing, and tanned apparently in some special sort of smoke. So many young ducks were brought in, both alive and dead, that we had to call a halt on them. Ookpik, John Ell's daughter, brought me some snowy owl's claws she had been saving for amulets.

.   .   .   .   .

Let me tell you about a young white-rumped sandpiper [1] I found running about at the edge of a sort of garbage pile back of one of the store-houses. If you are an ornithologist you will remember that the white-rumped sandpiper is not a rare bird during the period of migration in the eastern United States; but its nesting-grounds are so far removed from civilization that its summer habits are but little known. It was one of the most interesting species I was to study on Shugliak. When I saw this fine young sandpiper running about among barrel hoops and shavings and tomato cans I was annoyed. The bird would lose caste in such surroundings. I picked up a pebble, hit a tomato can near him with a loud

[1] White-rumped Sandpiper, *Pisobia fuscicollis*.

sock! and frightened him over to a mud-flat where he belonged.

．　　．　　．　　．　　．

The post hummed with life for two days after the call of the *Nascopie.* All the Eskimos had come in at "ship-time" to help with the loading and unloading. After their work was finished they gave themselves over to talking and singing and romping and loving. All of them, men and women and boys and girls, played a sort of ball-game, chasing this way and that among the boulders, shouting hilariously. I never learned just what the point of the game was. Apparently the women were to run after the ball, acting as if they wanted to catch it or pick it up, then screaming and giggling as they decided not to touch it after all. There must have been some penalty for touching the ball. They played the game so furiously I could not help wondering if penalties would be paid all winter long.

The boys had also a football, and their game consisted in hitting one another as hard as they could with the ball.

On the third day after "ship-time," the post became silent. Tents were pulled down and rolled up and bundled into motor-boats. Dogs and children were assembled on the gravel beach, waiting to be packed in when the proper moment arrived. I could see Sam and Jack shaking hands with the Eskimos and saying *"Tugvahootit."* [1] Not one of them came to say *"Tugvahootit"* to me, and I was a little shy about going down to the beach to shake hands with them. I didn't want them to think I was forcing myself on them. You see, I really wanted to like them, and to have them like me. I wasn't after being taken into their tribe, or being shown the mysteries, or being invited to their weddings, or having wives

[1] "Good-bye."

offered me, or anything of the sort. I just wanted to like them and to have them like me.

After they had gone, Jack and I had a talk. He said to me, in his quaint tongue (I can hear him saying it now; it was an important thing he was saying): "The people likes you, Doc. They tells me just now they's going to help you get all sorts of birds while you're here." Good boy, that Jack. You don't find many as full of imagination as he. Telling me that just when it was the thing I most wanted to hear. You'd think he had been through just what I was going through, this difficult shyness and uncertainty. The Eskimos all loved him because he was fearless and unselfish.

. . . . .

One afternoon we gave two of the useless, grass-stuffed duck-skins to the dogs. There was a grand fight. What a rush for those foul-smelling, sad-looking "specimens"! What a tearing and a pulling and a jerking! Puppies got drawn into the fray, just as puppies always are being drawn into frays, and they were bitten and nipped and pummelled along with the rest. The ducks were torn to shreds and more or less swallowed, stuffing and all. The feast (or the fight; call it what you please) ended in a five-minute chorus of howling. It was a chorus wonderful to hear.

. . . . .

The Eskimo women were to make me some sealskin *komik*.[1] The first pair were about two inches too short. The second pair also were too short. The third and fourth pairs I could get on, but they hurt my toes. And I wore them out in a week because I stumbled around so. The women had a

[1] Knee-length boots, the upper part made of *netchek*-sealskin with the hair left on, the lower part of *oogjook*-sealskin with the hair scraped off.

hard time realizing how big my feet were. For a while I decided to give up "going Eskimo," and wore tennis shoes.

.   .   .   .   .

You are wondering about this tale you have heard of Eskimo women chewing-down skins to make them soft. You make a face when you think about it. "Dirty people," you say, "chewing away at the filthy stuff. Must be awful!" Of course the Eskimo women chew-down skin. Whenever they sew caribou-skin or sealskin or polar bear-skin they chew-down the edges to make them pliable. They chew-down the stiff, dry *komik* of the hunters almost every day. But there's nothing so very repulsive about the matter. They don't go at it as if they were going to be talked about, the rest of the world over, for doing it. The kindliness of their faces, the efficiency of each movement gives a dignity to their chewing. It is their task, a task they can do uncommonly well; and they do it cheerfully. I confess it does give you a queer feeling at first to see pretty women like Pumyook's wife chewing away at boots with her dazzling teeth. You would consider Pumyook's wife pretty no matter where you might encounter her. You would be bound to think her teeth too handsome for chewing *komik*. But you might take the bright smile from her face by telling her there are certain tasks pretty Eskimo girls should not do. Better that you tell yourself something. Something like this: "Isn't she pretty? The very way she chews-down the *komik* is pretty. Her gleaming, clean teeth are so very white against the darkness of the sealhide!"

.   .   .   .   .

Near the post I continued to hear a strange cry from the lakes, a cry like the yelp of a puppy being kicked or clouted with a piece of fire-wood. I learned this sound was given by a Pacific loon [1] just before it dives.

．　　．　　．　　．　　．

On the twenty-fourth of August I caught a little dull-yellow butterfly, and saw some small pink flowers among the brown grass. Summer was almost done.

[1] *Gavia arctica pacifica,* an American subspecies of the well-known Black-throated Loon of the Old World.

CHAPTER IV

Prairie Point

THE twenty-fifth of August was one of those glorious skyless days. One of those days when you look up and wonder what color the sky is, or if there is a sky at all, or if it isn't just nothing at all above you. On the twenty-fifth of August, Sam and Jack and old Angoti Marik and I got into the motorboat called the *Robert Kindersley* and tripped across Coral Inlet to Prairie Point.

You are saying to yourself: "What manner of Arctic is this? Motor-boat indeed! Why not a *kayak* or an *oomiak? Kayak* and *oomiak* sound much more like the Arctic!"

As a matter of fact Shugliak's Eskimos of today have no *kayak* or skin canoe, such as the Labrador Eskimos have. Nor do they have any craft that may properly be called an *oomiak* or woman's boat. The nearest they have to a *kayak* is a canvas canoe, the sort you yourself go canoeing in, a canoe that is brought to them on the *Nascopie;* and the nearest they have to an *oomiak* is a motor-boat. I confess it gave me a jolt to find myself among *kayak*less and *oomiak*less Eskimos, especially when some of these same Eskimos were wearing pale blue suspenders and creaking tan shoes. But I was to find that pale blue suspenders and tan shoes and canoes and motor-boats have little to do, after all, with this matter of being, or of not-being, an Eskimo.

We got to Prairie Point in an hour or so. The Inlet was calm and, like the sky, almost without color. Out where sky and water met there was no horizon line. It was as if there were a great hole in the world through which we looked out into luminous space.

Prairie Point was a sort of low peninsula, nine-tenths lakes, seven miles from the post. The beach was of angular, glaring white chunks of limestone, difficult to walk upon.

Birds were everywhere: thousands upon thousands of Arctic terns, some of them feeding their flightless, downy young; myriads of red phalaropes bobbing lightly about on the water, twirling this way and that, stirring up the mud with their lobate feet; rafts of young old-squaw and king eider ducks, churning the surface as they made their way to a safe retreat or diving with a splash when close-pressed; solemn pairs or foursomes of parasitic and pomarine jaegers, dashing down among the terns, intent upon robbing them of shrimps or fish; rosy-breasted Sabine's gulls circling high in air, the jet-black tips of their wings seemingly cut off from the rest of their bodies and floating about strangely so always in place alongside the birds.

We had tea there on the beach, full in the sun. The whiteness of the limestone was almost intolerable. The terns scolded us incessantly.

Jack and I went inland. After going about three miles we separated. I was after jaegers. The mosquitoes were a little troublesome, though the mid-summer "fly season" was past. All at once I heard a shouting and turned to see Jack standing in the middle of a huge lake, all his clothes off, a gorgeous purple something tied about his loins, chasing a family of brant geese. He plunged this way and that through the shallow water, waving his arms and shouting and throwing rocks. Now he was calling me. Loaded down with gun and camera and bird specimens and that floppy hunting coat of mine, I ran as best I could toward him, envying him his shining nakedness, and that easy type of weapon he was using. When I reached the edge of the lake he had already killed two young geese.

I shot the parent brant without trouble, for they were in their mid-summer moult and could not fly. Finally I thought we had the remaining young cornered. I was all for getting a photograph of them. They swam to the opposite edge of the little cove into which we drove them, and waddled out onto the shore. I fancied we should have no trouble in catching them now. In fifteen seconds I changed my mind completely. The scrawny goslings lifted their stumpy, featherless wings, made a comment or two, and tore across the tundra at a speed to make a fox blink. Of course we never caught them. They reached another lake, promptly dived under, and didn't show themselves again until they were a long way from shore.

Funny, slim Jack, and that outrageous purple silk scarf, or handkerchief—or what was it anyway?

"It's a bathing suit, Doc. The people likes them here. You can have one, too. I'll get you a red one!" And I did get one; only it was orange-colored, not red.

When we got back to the post we learned that John Ell had been seal hunting.[1] It had been a grand day for *netchek-*seals,[2] for the bay had been so calm, and he had got one. I watched him skin the carcass. He used a great, broad knife, ripping the skin from stem to stern and pulling it back rapidly. There wasn't much blubber, for it was summer, but there were gallons of blood and the odor was noticeable. The dogs sat round in a big circle, their faces wrinkled and twitching in that funny, wolfish, Husky-dog fashion, their ears

[1] The firearm most commonly used by the Southampton Eskimos is the Savage high-power .22 rifle. This rifle is offered by the Hudson's Bay Company in exchange for white fox skins. The Eskimos have very few shot-guns and no revolvers.
[2] Ringed Seals, *Phoca hispida*.

lifted and put forward, and their avid tongues licking their drooling lips.

John cut the seal up, snipping through the ribs as if they'd been only tender stalks of celery, and laying the soggy, dark red chunks on a big, rusty, tin platter. Then the women came for the meat, leaving the platterful of blood for the dogs. There was a center rush such as our better football captains never dreamed of, a wild lapping of tongues, a guttural threat or two, and a glorious free-for-all. Long teeth sank into eyesockets and ground down hard. Cowards rushed for the water, where they stood sadly watching the fray. The Eskimo women burst out with chunks of wood and cracked the noses of the most savage brutes; but the fight continued. The puppies, as always, found themselves engulfed in a rapids and whirlpool of snarling fury, and yelped and whined pitiably. It is thus a Husky pup becomes a Husky dog.

You might, after you have lived for a year on Shugliak, wish to be an Eskimo. But you would not wish to be a Husky dog.

CHAPTER V

Cape Low

THE chances are you have never heard of Cape Low. It is a flat, rounded point of land at the southwestern corner of Shugliak. It is wild country. When you stand there looking out into the blue-grayness that is Hudson Bay or Sir Thomas Roe's Welcome or the Bay of God's Mercy, or into the yellow-brownness that is the vast interior of the island, you have a feeling that you are a long way from home.

Late in August, Sam Ford and John Ell and I left the post by motor-boat, headed westward. I wished to verify, if possible, some of the reports concerning nesting grounds of blue geese at Capes Low and Kendall.

We made about fifty miles the first day, passing one of the "peculiar shedlike hills" of the charts, some cavern-filled limestone bluffs, an ice-bank that had the appearance of a glacier of some sort, and the mouths of several streams. The weather was good. In the deeper waters we saw many *netchek*-seals. They had a way of coming up near the boat, turning their doglike heads about, rolling their big eyes sadly, and lolling this way and that as if they were at the point of taking a nap. If we whistled they sometimes lifted themselves well out of the water, then sank back down and disappeared.

All the coast of this part of the island was very low and the water shoal, so we had to steer our course several miles out. We put to shore that night at a place I called Four

Rivers. A big white stone that we thought was a polar bear marks the place, if ever you wish to visit our campsite. We propped the motor-boat up with two-by-fours, for the tide was going out. Everywhere about us were heaps of stinking seaweed, filthy, rotten masses of the stuff two or three feet deep. Shore-birds swarmed about this kelp, running this way and that, probing and foraging.

In the far distance rose another of the "peculiar shedlike hills" that John told us the Eskimos called *noovoodlik*.[1] We had quite a discussion about these hills, and I decided they must be some sort of glacial deposit, perhaps terminal moraines. At any rate they were huge heaps of gravel rather than solid masses of rock.

I couldn't help calling this place Four Rivers because there were so many streams in the vicinity. My personal feeling is that there were about forty streams (for I was always crossing one, no matter where I went), but conservatism is said to be the mark of a good explorer, so I boiled the forty down to four. They were shallow streams with clean, sparkling, gravelly bottoms, and they wandered about everywhere through the grassy meadows, and among the gravel mounds, as if they might eventually lead to some lake in the interior rather than out to sea.

I walked inland a considerable distance, heading straight for the lofty, lugubrious *noovoodlik*. Evening was falling. The horizon was hung with clouds. There was a mistiness over the tundra. All at once, from the far distance, I heard a mellow trumpeting. Straining my eyes I made out two tall bird-forms stalking slowly across the meadows. Cranes! Magnificent creatures, the stately *tutteeghuk*[2] of the Eskimos. I walked toward them, but they took alarm when I was yet

[1] A kind of high, gravel mound or hill: a common, not a proper noun.
[2] Little Brown Cranes, *Grus canadensis.*

far away, and made off with square-tipped wings flapping stiffly.

The sandy soil was honeycombed with lemming burrows. Whenever I stood still I could hear the little mice gritting their teeth at me. The sound made me think of tiny skeletons rattling. In the damp gravel I made out the tracks of a band of caribou (there must have been some young animals among them) and the trail of an Arctic fox. Circling over-

ISHOONGUK: LONG-TAILED JAEGERS

head were several jaegers. I saw two of them chase a Lapland longspur high into the sky; bite it in mid-air, pulling all its tail feathers out; follow it to the ground and kill it, there to swallow it in a gulp or two.

We resumed our journey early next morning, taking advantage of high tide. All went well until noon, when the wind rose and we found we were making no headway. The water became greenish-yellow as the waves stirred up the sandy bottom. There was no rain, but there was fog, and the spray from the wave-crests drenched us every time our boat nosed through them. We flopped and whacked and lunged about so wildly that we decided to make for shore, though

Sam and John both said there was no decent harbor anywhere in the vicinity.

I knew I was becoming seasick. I stayed at the wheel as long as I could, hoping the fresh air and the lashing of water across my face would keep me fit. But I went under. The sickness itself was bad enough, but the torment of it was the feeling of utter uselessness that had me in its clutches. I would try to rise, believing that a show of interest would do me good, then sink back convinced that the world would somehow have to get on without me. I had never been badly seasick before.

Sam and John were good seamen. Faces and oilskins wet, they carried on. They weren't merry. Their faces were stern. They talked to each other in Eskimo; but even if they had talked in an understandable tongue I couldn't have caught the point of much of their conversation.

We were slapped about so roughly it seemed we must surely break apart. As we slowly drew nearer shore we began thumping the bottom. This new sound, this new sensation didn't seem so different to me from the various sounds and sensations I had been experiencing for the past hour; but I could see it was different when I looked at my companions. Then we hit a big rock. There was a roaring of the engine, and a silence save for the hissing of the water and the thrashing of the wave-crests across us. John started the engine again. There was a wild roar, but we did not move forward. The propeller was done.

The next two hours are not a happy memory. Try as I might and did, to be up and about I could not be anything but sick. When I tried to rise I simply fell over in a heap. I got up once and hit my head so violently on some corner of the little cabin that I thought I'd better stay put, rather than add blood to the confusion about me.

Sam and John were unlashing the little white tender. We were bumping the bottom a good deal now, but it was soft bumping. The rowboat was finally dropped alongside. Sam let himself in. The anchor was thrown in near his feet. He somehow rowed ahead. It was a man's job, rowing through those waves, carrying forward that anchor with its heavy chain, managing somehow to drop it some distance ahead, then rowing back to help John pull on the chain and thus slowly move us shoreward. I don't know how many trips Sam made. It was enervating, exhausting work. I despised myself for being such a piker.

Finally I was able to stand and helped shove the boat forward, using a long boat-hook. The waves were not so fierce now. I had to be on guard every second lest I fall overboard. I may have helped a little—not much.

It was dark when we finally came into the shelter of a low-lying point of land. Once more the motor-boat was propped up with two-by-fours. On the shore an Arctic fox barked merrily as Sam lighted the lanterns.

In the morning the boat was white with frost, and the mud of the shore firmly frozen. When I went for fresh water for tea I saw a family of cranes feeding near some old stone fox-traps. Everywhere along the shore were bones of caribou and bear and walrus. This was a campsite of the extinct Shugliamiut.

I caught a half-grown whistling swan that day after a long chase into the interior.

We found, to our great delight, that the propeller had not been broken off. John mended it, making a new pin with a spike he found somewhere. We were ready to go ahead, but the barometer was low and the wind blowing hard.

We stayed at this place two days more, living in a tent. I had a busy time collecting and preparing bird specimens, which I kept in the damp, cold hatches under the cabin bunks. We caught a crippled old-squaw duck, a lively bird that may have been bitten by a jaeger when it was young. We kept this bird in a little box in the tent. It wakened me in the morning by running over my face with its big, flappity feet, and nibbling at my nose, lips, and moustache with its stubby bill.

Once more we pulled up camp and headed for Cape Low. It was not good weather, the reefs were ugly, and the sky forbidding, but we went on, and finally reached calm water at the mouth of *Kashigiaksoak* (Kashigiak River), a fair-sized stream. Here, on a gravel-bar, a herd of *kashigiak*-seals [1] were basking in the intermittent sunlight. We anchored at the mouth, in the shelter of a ten-foot bank. A seal was at play not far from us. John got out his rifle and hot-footed in pursuit. He wounded the animal but did not get it. Then he began a long stalking of the herd on the gravel-bar. We heard a distant shot. He had got a young one. I helped him drag the animal in. It was a beautiful creature, much more handsomely marked than the common *netchek*. I learned that the skin of the *kashigiak*-seal is highly prized by the Eskimos. They like to make their sleeping bag covers and their more fancy clothing from it. We dined on *kashigiak*-seal that night. The boiled meat was tough and it had a strong flavor, but we were hungry and ate a great deal of it. I was not yet enough of an Eskimo to try any of it raw.

On the next day I walked out to the end of the cape. The beach was one of the most beautiful I ever saw. It was very smooth. The limestone particles and sand at the water's edge were fine and white. Farther up the slope the gravel was coarser and darker, and the rim at the top was composed of great chunks of limestone, so neatly arranged you would

[1] Ranger or Harbor Seals, *Phoca vitulina concolor.*

think someone had been making a gigantic rock-garden. I came upon some old casks—the only signs we found of Captain Murray's or Captain Comer's whaling stations. I got a good many birds that day. It was not easy walking with those two fifteen-pound swans swung with cords from my shoulders, a brant goose and a red-throated loon hung from my belt, and four pomarine jaegers, a parasitic jaeger, a young red-throated loon, a big Pacific loon and a Hudsonian godwit in my collecting-basket or in my hands. The next day I had to put in preserving this lot of material. And we dined on swans' livers and gizzards.

We had hoped to journey farther, on to Cape Kendall. But the weather was so disagreeable we decided it sane not to try to cross the Bay of God's Mercy. We searched for a nesting ground of the blue geese, but found only feathers and droppings where the birds had been feeding and resting and preening their plumage. We saw huge flocks of both blue geese and lesser snow geese, and watched these birds starting on their autumnal flight to James Bay and the mouth of the Mississippi. John told us he had found both *khavik* [1] and *khanguk* [2] nesting in great numbers near Cape Low several years before.

On September 4, we started for "home." It was cloudy, windy, chilly, and raining. We journeyed about twenty-two miles, and decided to make for shore near an Okomiut encampment. Once more the kelp, this time acres of it. The stench was frightful.

We had to stay here five days, waiting for good weather. The Okomiut were living in shabby little *tupek* near the shore. They didn't have much to eat, though they had suc-

---

[1] Blue Geese, *Chen cœrulescens.*
[2] Lesser Snow Geese, *Chen hyperborea.*

ceeded in getting two *oogjook*-seals [1] and three polar bears. I visited the camp several times, trying to be friendly. Dogs lay about everywhere waiting for some excitement to turn up or for a fish-head or old boot to be thrown out. A raven flew down into camp one day and the dogs chased it savagely, snapping at it with their vicious jaws.

One day I saw some parasitic jaegers chasing a sandpiper. The little bird flew far out to sea then headed inland, coming straight for me. Just as the big pirates were closing in he dived into a crevice in a pile of stones. The jaegers were after him in a minute, alighted on the stones and peered this way and that into the crevices. I came up and frightened the hawk-gulls away. Lifting stone after stone I finally found the sandpiper crouched in a lemming burrow, breathing rapidly. I held the bird in my hand, opened my fingers and bid him be more discreet about jaegers from this time on. He did not leave at once, but lay there looking at me with his bright eyes. When he realized he was free he stood up, gave a short call, and dashed off, blithely as if the world held not a single enemy. Sandpipers must have very poor memories; or maybe the excitement of being held in a man's hand is greater than that of being chased by a jaeger.

Sam told me that the kelp beds at Cape Kendall were terrible, and that one of the Aivilik Eskimos had once almost suffocated in them.

The Okomiut had a feast on polar bear. They ate it both raw and cooked, and there was a gorging. They had an outdoor fire made of planks from an old motor-boat, and there was shouting and dancing and playing of ball. On the following day there was much sitting around and much sleeping.

The morning we started for the post there was heavy

[1] Bearded Seals or Ground Seals, *Erignathus barbatus*.

frost, and a scum of ice on the salt water. We encountered fog, but the sky finally cleared and we made our way through South Bay in the glassiest water. We saw a herd of *kiolik*-seals,[1] leaping and bounding along like dolphins or porpoises. We were glad to be back at the post. I needed the refining influence of a mirror.

That evening I skinned birds and rewrapped specimens until long past midnight. When I came outdoors before going to bed, I stepped on one dog and kicked another. The dogs took their spite out on each other instead of me, and a fight started. I clouted every dog I could see, and a howling commenced. From inside the house I could hear shouting: *"Palaghit, palaghit!"* [2] and I knew the post had been wakened. Sam got up and came out to make certain nothing was wrong. We had a cup of tea together. Then I went to bed, taking with me a cake of lavender soap, something I could smell, thereby forgetting the beds of stinking kelp.

[1] Greenland or Harp Seals, *Phoca grœnlandica.*
[2] "Keep quiet! Shut up!"

Chapter VI

Tundra Magic

Just east of the post was a wet meadow, almost a marsh, fringed with a sort of sedge that Sam called bog-cotton. Beyond this meadow a granite ridge half a mile long rose abruptly to a height of about thirty feet. From the post the ridge had a dark, colorless appearance. But when you examined the rocks carefully you found them to be covered with delicate lichens of many colors: gray-blue, pale green, orange-red, dusty yellow, and brown.

It was snowing lightly, now that it was September. The wind tossed the fragile flakes about playfully, tiring of them, letting them drop into the grass of the lowlands, into the clumps of willow, into the sheltered niches among the rocks. By evening there was a fluffy heap all along the base of the ridge and so much whiteness on the marsh that at a little distance you could not make out the waving white tassels of the bog-cotton.

I walked along the ridge, thinking of the winter. The wind had kept bare the most exposed places, and here the dry stems of the short weeds, the fine, curled grasses, and the cleft pods of legumes rustled and rattled.

I noted ahead of me a patch of snow, not as large as my hand, that, oddly enough, had defied the wind and clung to the very top of a rock. The mosses and epiphytes near this snow-patch were, perhaps, a little regular in pattern and there was a touch of red somewhere, not quite the red of a lichen or even a berry. Was there, near this patch of peculiar red, a bead of glistening black?

Hardly conscious that I had asked myself why patches of September snow should occasionally defy a strong wind, I walked closer . . . and a ptarmigan rose on neatly booted feet, craned his long neck, jerked his tail, and sauntered up the rock. The bead of glistening black was this ptarmigan's eye, the patch of peculiar red his naked brow!

How completely that bird had fooled me—not he himself, of course, for the ptarmigan probably didn't care whether I saw him or not—but that perfect color pattern of his! A patch of white winter plumage nestled among the grays and browns of his summer coat, just as the new snow-drift nestled between the base of the ridge and the brown marsh.

The ptarmigan was almost insolent. He walked out of the way merely to avoid having his tail trod upon. Obviously, being larger than a weasel or fox, and "not half so spry," I could hardly be an enemy and was therefore not worth watching. Why he craned his neck or flicked his tail is beyond me: calisthenics of a sort, perhaps; or an involuntary response to some dim trace of suspicion. So accustomed was he to being undetected that he paid no attention to pebbles thrown at him, even though these whirled but a few inches past his head. If in walking about him, I took a course that promised him no physical discomfort, he remained motionless, eyeing me drowsily; or, to state the matter more accurately, including me in his looking simply because he was looking in my direction.

The ptarmigan was not, in fact, much interested in me. My attitude toward the ptarmigan, on the other hand, was one of amazement. His plump form was neat in contour. The tiny vermilion comb above his eye set off the surrounding sombreness of moss and dry grass to perfection. His feet, feathered to the very claws, were lifted and set down upon the lichens with gentleness and unstudied precision. I marvelled that any single biological law, or complex set of laws, could account for a color pattern as handsome as that of his neck, back, and rump; and that, at precisely this season of the year, he should carry about on him his own little snowdrift, so that, whenever he chose, he could squat on the ground, begin to feel like any chance piece of old granite, look stonily at the world, in a manner of speaking, and all at once—well, simply cease to be a ptarmigan!

How fortunate, thought I, to come upon a ptarmigan so near the post! And a rock ptarmigan,[1] too, the smaller and rarer of the two species found on Shugliak. What a sketch I would make—a sketch that would record that impertinent, lazy twinkle of eye, that reluctance to move one step out of the way, that plump squatness that might, without the remotest influence of any snake-haired Gorgon, on the instant turn to stone!

"Pe-ar, pee-ar!" came a gentle voice almost underfoot. There, about four feet away, its tail jerking half-acknowledgment of my presence, was another ptarmigan, another lichen-covered boulder with its little snow-drift. By the time I had crossed the ridge, bound for the salt-water "lake" beyond, I had encountered nine of the birds, apparently a family gathering of some sort, and I may not have met them all.

[1] *Lagopus rupestris.*

"Smug creatures!" I thought. "Chance possessors of the magic cloak of invisibility! You are nicely equipped indeed, for your Arctic existence. But you act so much like a chunk of rock that you have taken on, unawares, the soul of a rock. And you have the brain of a snow-drift. You have warm moccasins for your feet and leggings to match. You are perfectly comfortable after a meal of willow-twigs. Wind and storm you do not fear. Cold weather is your delight. Look at you, sitting there like barnyard chickens in the middle of this wild country. You are so content you do not even appreciate your good fortune, and you have always been so lucky that you probably never even thought about such a thing as intellect."

As I turned to leave the statuesque birds perched here and there on the rocks, I noticed, not far away, some feathers waving from a willow bush. Some of these proved to be patterned in gray, brown, and black; others were pure white. A white wing, its strong, curved, flight feathers stained with blood, lay to one side; to the bones clung bits of flesh, still damp. Here a ptarmigan, even he of the invisible cloak, had been struck to the ground by a gyrfalcon. He had moved at the wrong time. Tundra magic, this time, had failed. For there were eyes in this gray-brown world far sharper than mine.

CHAPTER VII

Native Point

AUTUMN is a brief season on Shugliak, if indeed there is an autumn. Snow is likely to fall at any time of the year; the grass is more or less brown and withered all summer; the moss and lichens are gray or gray-green or pale blue whether it be Summer or Winter; the yellow buds of low-growing poppies scarcely open before they have become empty pod-heads; the leaves of the dwarf trees scarcely are of full size before they are turning yellow and scarlet; and the south-ward migration of sandpipers and plovers begins long before all the young birds of the tundra are full-feathered or even safely out of their nests. You think it is summer for you see a pink flower in the moss, or because you hear a snatch of bunting-song; but on the same day you notice that the Arc-tic hare's face and ears are turning white and that the weasel is taking on that strange parti-colored appearance—and you decide it's Autumn.

The Eskimos think of Autumn as two seasons. They have an early Autumn that they call *Ookiukshak,* the "time of the first snows and the migrating birds"; and they have a late Autumn: *Ookiak,* the "time when the islands in the bays freeze shut."

We had a considerable snowfall on September 11, enough to make the landscape white, so *Ookiukshak* must have been upon us. Snow buntings and longspurs were still with us, however, and red leaves still clung to the six-inch sprays of birch.

I kept hearing about a place called Native Point.[1] "It's the place where the Old People died out," Jack explained. "They's lots of old *rooins* there, and graves, and bones everywhere. The people lives there now, only they lives in other sorts of houses now. Honest, Doc, they's lots of graves there and bones everywhere!"

When Jack spoke of the "Old People" he always meant the tribe of Eskimos that originally inhabited the island, the real natives of Shugliak, the Shugliamiut. It was at Native Point, then, that the Shugliamiut had had their last stronghold. It was at Native Point that the strange and fatal epidemic had struck them, wiping them from the face of the earth. I longed to see the place.

On September 18, Jack, John Ell, Curly Joe (an Aivilik whose real name was Kayakjuak) and I boarded the motorboat *Shookak* bound for Seahorse Point. The purpose of our trip was twofold. I was to study and collect birds and mammals and see as much of the country as I could. And John Ell was to locate the carcass of a whale he had mortally wounded the preceding July and bring back what he could of the valuable baleen, or whalebone. This going after the carcass of a mortally wounded whale seemed strange business to me. I knew full well that a whale was a big mammal, but I didn't think it big enough to be found anywhere after the lapse of several months in any such vast country as this. I made no comment, however.

We left the post in a wild snow flurry. Our first stop was to be Native Point, a sort of peninsula about thirty-five miles to the southeast.

En route we passed Bear Island. We didn't see any bears,

[1] Tunirmiut on some charts.

and I decided the island ought to be called Bare Island. But the Eskimos perceived no reason for smiling over this decision.

When we got to Native Point we were all glad to walk a bit, for the sea had been rough. Here the Eskimos' *tupek* of canvas and sealskin were pitched not far from the water's edge along a little cove, and in the very midst of the remains of the permanent stone houses of the extinct Shugliamiut. Dogs lay everywhere, in the shelter of *komatik*,[1] close up against the *tupek*, near the big boulders. I wondered why these dogs should be so inert, but ceased to wonder when I had occasion to look at, to walk round and over, and to smell the several half-eaten, crudely skinned carcasses of white porpoises or *kellilughak*[2] that lay here and there all about the encampment. The dogs were too full to move. One of them was kicked for lying in a thoroughfare, as Husky dogs are wont to be kicked for lying in thoroughfares, and when he yelped he disgorged a mass of putrid whale flesh. It was a somewhat gruesome place: these ruins of an extinct civilization; these rotting remains of monsters from the deep; these matted-haired gourmands of dogs lying about too full to move.

The Eskimos themselves also had eaten well. Most of them were lying down or sleeping. We roused some of them, bidding them be ready to come with us, on the morrow, to Seahorse Point.

That afternoon I had my first meal of *muckluck*. *Muckluck* is an Eskimo delicacy. It is the outermost covering of the white porpoise's skin, a somewhat gelatinous substance resembling in appearance and somewhat in flavor, boiled white of duck's egg. Either I was not very hungry or I

[1] Sledges.
[2] White Porpoises or White Whales, *Delphinapterus leucas*.

actually did not take a liking to *muckluck,* for I did not eat much of it.

Jack and I looked over the ruins of the "Old People's" houses. Evidently the houses had been built of stone, then roofed over with strips of whalebone and some sort of coarse hide, probably walrus hide. The ruins stood, for the most part, on a narrow flat in the shelter of a turfy hill. In the houses were all sorts of chambers or closets or storage-bins that had been built up of smaller stones, some of them lining the bases of the walls in the manner of hollow seats. A little stream ran through one of the houses, sparkling in at one side and sparkling out the other. Had the house actually been built thus, over the stream? Who can tell?

Lying all about the dead village were pieces of caribou bone and antler; walrus, seal, and bear skulls; remains of old sledges and harpoons and various other implements. We touched none of these, for I was no ethnologist. I thrilled as I pictured such a person as Dr. Franz Boas standing where I was standing, with all these rich remains of an interesting culture heaped about and waiting to be studied by an expert.

Leaving the ruins of the houses, we made for a high gravel hill about two miles inland, passing on the way heaps of walrus and caribou bones and whale ribs and vertebræ. When we reached the top of the hill we found ourselves in a cemetery—graves built up of flat stones, some of them so exposed that we could plainly see the remains of corpses lying there. Here was the grave of a child, the skull lying on its side, the leg and arm bones all doubled up as if the body had been buried in the smallest possible space. Here was the covered grave of an adult. We could make out the remains by peering between the stones. Some of the graves had obviously been ravaged by dogs or wolves, for the stones had been strewn about hither and yon and human bones were scattered everywhere. I did not see any harpoons or bows and arrows or trinkets, but I did not carefully examine any of the graves, and these objects may have been buried in the gravel or under the bones.

We went to the farther end of the hill and took a look at the vast interior of the island. How endlessly the tundra stretched off to the north and east, off to the rugged Porsild Mountains and the cliffs along Fox Channel, off to the high

UGHIK: OLD-SQUAW DUCKS

country about Mount Minto! Then we went down the hill and chased *ughik*-ducks [1] about the lakes. They were still moulting their wing feathers even at this late date.

Jack explained to me that walrus were killed in considerable numbers every summer in the bay called Big Rock Bay, or Shallow Bay, or Native Point Bay, that lay just to the north of us; and that the Eskimos, after killing the great brutes, simply let them drift to shore where they later went with their dog-teams to find and cache the carcasses.

As we walked along the shore we saw a dog-team approaching from the north. Here they came, thirteen dogs and a

[1] Old-squaw Ducks, *Harelda hyemalis.*

big sledge upon which sat two chubby, well-fed Eskimos, all heading for the encampment at the point.

"Jump on, Doc!" shouted Jack.

"Do you ride these things in summer?" I asked, forgetting that we had had a snow-flurry and that the season called *Ookiukshak* was upon us.

"Sure, the people rides them any time. The dogs pulls you anywhere if you beats them hard enough!"

And I got on, somehow, for my first *komatik* ride. I sensed in an instant that my legs were too long for this sort of vehicle, and wished as we thumped and bounded along, that I had a cushion or at least an old sack under me.

The dogs whined a good deal and looked round as if for sympathy. The walrus-hide whip cracked, flicking out a neat chunk of ear. There was a yelp of pain, much panting and much whining. The dogs were telling us how hard they were working. They were pulling us directly over gravel and mud and moss most of the time; the going was much easier, even on an up-grade, when we struck a snow-bank.

We began the ascent of a small, bare hill. One Eskimo got off as the dogs strained and whined and bent low their backs. I got off too, telling Jack that I would walk to the encampment, looking for specimens on the way.

Afoot once more, I mused upon the tragic story I had heard about the ancient Shugliamiut. They had been an intelligent and resourceful race. Theirs had been rather a unique civilization. I recalled that funny picture, in Captain George Francis Lyon's account of his journeyings, of an Eskimo called "a native of Southampton Island" sitting so placidly on a sort of raft of inflated sealskins. I recalled too the brief but highly interesting notes pertaining to this people in Captain George Comer's old diaries; and the scholarly discussions of them in Therkel Mathiassen's reports. The Shugliamiut had been an ocean-loving, ocean-hunting people. They had lived on seals, walrus, and fish, and had killed even the largest whales, the blubber and meat of which they used as food; and the baleen and ribs as construction material for their permanent houses. In these winter houses, so I had heard, two to four families lived together as a rule.

John Ell told us that the Shugliamiut knew how to build *igloo* or snow-houses,[1] but that such houses were built only as temporary camping quarters.

And Muckik told us of little gull-catching houses they had built, little *igloo* with thin roofs. A man would get inside such a house after putting a piece of fish on the roof. Then, when the gull came to eat the fish, the man would reach out through the roof and grab the bird by its feet.

That evening I sat with the Eskimos for a long time, listening to their conversation about whales and bears and dogs, and to their tales, none of which I understood. I watched their interesting faces in the dim light of the candles, *koodilik*-lamps,[2] and lanterns, wondering how soon they would learn not to be afraid of me.

It was on that evening that I began to sense how different the two tribes now living on Shugliak actually were. Not long ago I had been in a camp of the Okomiut, the Baffin Islanders, over near Cape Low. Now I was among the Aivilikmiut, the walrus-hunting people from Repulse Bay. I seemed to like the Aiviliks better. Was it that they were less furtive, more open-faced and genial, more likable? Or was it simply that I had gradually become more at ease myself and was

[1] An *igloo* is a snow-house, not an ice-house.
[2] Seal-lamps: stone dishes in which seal-oil is burned.

therefore giving the Eskimos a better chance to be themselves?

But Jack said: "Doc, I likes the Aiviliks a lot better than I likes the Baffin Landers. They's cleaner and more friendly. Doc, I just loves the Aiviliks sometimes!"

Chapter VIII

A Dead Whale

When we left Native Point next morning we took Kooshooak, Muckik's son, with us. Kooshooak was, I should say, about twenty years old.

We now began looking for John Ell's mortally wounded whale. I asked a few questions about how you look for such whales but got no satisfactory reply, so contented myself with sitting back and smiling inwardly. Mortally wounded whale indeed!

We drank some coffee. I have no idea what was wrong with the stuff (I did not notice what date was marked on the can) but no sooner had we downed it than we all began to notice how smartly we were rolling along through the green waves. Even the Eskimos' faces took on a peculiar, vacuous stare, eyes looking straight ahead but seeing nothing in particular.

We should all have been seasick in one manner or another in a quarter of an hour had it not been for the walrus. I was gazing steadfastly into the water alongside the boat, marvelling in a vague way at that slipping, endlessly varying, endlessly the same glassiness of surface through which we were moving, trying to fasten my thoughts upon anything at all but coffee, when suddenly there rose, not more than thirty feet from my very face, a monstrous gray-green head with two gleaming tusks and a dripping broom of whiskers. Clemenceau himself! It was like a dream. For a second or

47

two I honestly couldn't believe my senses. But there he was, big as life, looking at me from behind his whiskers.

"Aiviuk!" [1] I shouted, proving I was going Eskimo. The word *walrus* never even crossed my mind.

You never saw such a snapping out of it in your life. The crew of the *Shookak* came to as if they'd been stuck with hat-pins: Eskimos who had ten seconds before been mere representations of living beings leaped this way and that, hissing *"Aiviuk!"* between their teeth, lunging for rifles lying in the most preposterous places, crawling over one another, stumbling this way and that, cracking their shins on boxes and edges of timbers.

This strange madness struck us like a bolt of lightning. There hadn't been the slightest symptom. Since I was the least affected, it was I who suddenly noted that no one was at the wheel, that the *Shookak* was charging and galloping round in circles, going at a terrific rate but going nowhere at all.

I moved over quickly as I could, found that all the *Shookak* really needed was someone to manage her a little and marvelled that these companions of mine could be in so many places in the boat at once. Jack was here now, peering out into the water for all the world like a starving hawk, eyes blazing; then all at once Jack wasn't here, it was John Ell instead, a killer now, no longer the smiling raconteur, no longer the genial companion. I steered on in spite of all the confusion, in spite of all this hopping over, this crawling across, this frenzied pushing and shoving.

Rifles cracked. But *Aiviuk* was far away now, headed for some distant clam bed.

Once more a calmness settled upon the *Shookak*. You could look at one of your companions long enough to tell it

[1] Atlantic Walrus, *Odobenus rosmarus*.

was he, not someone else. The course toward little Kikkuktowyak Island was resumed. But there was a brightness in those eight eyes about me. And no one thought of coffee any more.

We were to see more *aiviuk* that day, little bands of them swimming close together: great bulls with tusks as big as baseball bats; smaller cows, with shorter, slenderer tusks; and young animals that I suppose you could call calves (though it doesn't seem right to speak of a walrus calf), all rolling through the water, crowding and shouldering one another in their confusion, their eyes gleaming red as they blew through their coarse whiskers and made strange grunting, coughing sounds.

We ran through schools of white porpoises too, shy creatures that suddenly swarmed about us, giving us gleaming glimpses of their broad backs, blowing their tin-panny songs as plumelike jets of water shot up from their heads. So these were *kellilughak:* that gleaming whiteness out there that glided into being above the waves so gracefully and silently and as silently and gracefully glided away—that was *muckluck.* Incredible! A white porpoise in the sea is a creature of poise and beauty, of grace and mystery; beautiful and mysterious in its way as the aurora borealis, as a snow crystal, as the eerie harmonies of the dog chorus. But a dead white porpoise, hauled up on the beach, lying there among the kelp, its whiteness all battered and scarred, a heavy trace tied through its nose—such a thing is no white porpoise at all, it is just a big, flabby, repulsive carcass, all covered over with the stuff called *muckluck.*

We stopped that night in a little cove. The harbors along this, the southeastern coast of old Shugliak, were better than those in the Cape Low region. That is to say, there were harbors here; there weren't any harbors at Cape Low. A shallow stream ran into this cove. Inland about two miles was a crumbling limestone cliff about fifty feet high.

On the following morning we resumed our search for John's whale, steering our course close along the shore.

Countless thousands of old-squaw ducks rose ahead of us, shifting along the horizon like dark-colored vapors blown by a fierce wind. These birds must have been moulting, for we saw feathers everywhere about us floating on the water.

We saw more walrus, too, bands of from twelve to twenty animals, huddled closely in the water, sinking and rising, not making off rapidly as you might suppose. The sound of our motor-boat apparently confused them.

At about nine o'clock in the morning we found the whale. We were at breakfast "down below" when John shouted *"Akvik!"* [1] and everybody piled out for a look. There it was, a long way from us, just like a big black boulder on the beach. Above it gyrated a swarm of gulls and a few ravens. John had followed these birds to their feast.

As we approached the shore a raven circled over us. I shot the bird since I needed the specimen. As the black body struck the water John said something in Eskimo. His face was serious.

According to Jack, who explained the matter to me, the raven's falling to the left of our boat and its falling on its back had something to do with the success of our trip.

"Aren't you supposed to shoot a raven from a boat, is that it?" I asked, remembering full well how superstitious the Eskimos were reputed to be, especially about such creatures of omen as ravens.

"The people watches the ravens, Doc," Jack answered. "They tells what's going to happen by what the ravens does. You seen that one fall on its back, didn't you, Doc? Well,

---

[1] Bowhead, Right, or Greenland Whale, *Balæna mysticetus*.

that maybe means something. Maybe it means we'll see a bear or something."

We landed on the whale.

Walking on a whale that has been dead four months gives you strange sensations. Everything jiggles so. The whale is half-afloat, of course, but the floating isn't the strange part, it's the sinking down of your feet into that black, marshmallowy mass that doesn't give way exactly but feels as if it were going to. You simply gag at the odor, wondering that you never before smelled a really bad odor, and you marvel at the size of the brute. You've seen whale skeletons in museums, to be sure, and you've read all sorts of long descriptions with statements of measurements and so on, but here you are for the first time actually walking on a whale, a real whale that was killed in Hudson Bay; and there's Hudson Bay right there, lapping, lapping only a few feet away; and those persons that are shouting so, and talking so, and dancing little jigs way up at the other end—those persons are real Eskimos. What a continent of a carcass! What a stench! This wind's an ill wind no matter what it blows!

I jumped off to take a look. Here were the tiny eyes, way back in the middle of the body; here the vast tongue, big as a horse; here the strangely fringed whalebone, a whole cat-tail marsh of it, hanging in the mouth. I stepped off the width of the tail: twenty feet! I could imagine being whacked by a tail like that. Flip!—and Good-bye!

Everybody was happy. Half the precious whalebone was gone, rotted out and carried off by the waves, but a whole woodshed full of it was there and John was already at work sawing off those huge timbers of lower jaw-bone with his little saw.

Toward the rear of the gigantic torso there was a putrid cavern where the skin had been torn away and the viscera exposed. In the wet sand near by were the unmistakable footprints of a bear. Here the gulls swarmed, pulling greedily at the shreds of coarse muscle, fighting noisily with one another and getting out of the way whenever a raven came along. The air was alive with gulls.

A quarter of a mile away I had the good fortune to find all the missing whalebone buried in the sand.

"Tell the Eskimos I am glad I shot the raven, and that it hit the water with its back," I said to Jack.

And this strange and unexpected heresy having found expression, everybody looked mystified and dubious but not exactly unhappy. True, we had found the whale, and equally true, I had found the missing whalebone: I, the very person who had killed the raven that fell on its back in the water. Now, after all, what could be said to that?

John was sawing away at jaw-bone. He must have been a third of the way through one side by now. Kooshooak and Kayakjuak got into the little white rowboat and made off on a *netchek*-hunt. Jack and I took a walk inland.

We had walked about a mile when, with our binoculars, we sighted a polar bear. Attracted by the powerful odor of the whale, or perhaps, returning to his established dining-table, he was swimming strongly across the cove, unseen by any of the Eskimos and evidently unaware that his most dreaded foe was at hand. We decided that any attempt to signal to the Eskimos would only confuse them or frighten the bear, so retracing our steps to a rocky mound, we waited, watching.

The bear swam calmly shoreward. John continued his sawing and hacking at the whale. The gulls, traitorous gluttons that they were, knew that a feast was in the offing, so warned nobody. Just as the bear's forefeet struck bottom on a shoal near by, John suddenly stood up straight, then crouched.

Characteristically thoughtless and improvident, as an Eskimo nearly always is, he had left his rifle "somewhere else." Body bent low now, so the huge carcass would keep him from sight, he was scuttling down the beach, waving his arms in an attempt to signal the other men, then creeping back on all fours, to the whale. The bear appeared to sense that something was wrong, but came on, sniffing the air and turning his long neck this way and that.

We saw John pull a boat-hook from the sand and lash his great knife to one end with a strip of seal-hide. We gasped. He was improvising a harpoon for hand-to-hand encounter!

John had no chance to use this harpoon. An unexpected shot rang out. Another. The bear whirled toward the open sea, ducked his head and disappeared. Not far away the boat-load of *netchek*-hunters were rowing madly. There was another shot.

Presently we saw the boat coming toward shore, towing some heavy object. The bear had been shot and John was waving his boat-hook harpoon in air, shouting something in a loud voice.

When we got back we took a look at the animal. The Eskimos said it was about two years old. There was a long, detailed discussion of the shooting, illustrated with many and varied gestures. When the discussion was finished John said, a characteristic smile playing about his face, that he would have got that bear if somebody else hadn't made "so much noise" with a gun.

Chapter IX

Seahorse Point

SEAHORSE POINT is the southeasternmost corner of Shugliak. It was first seen by white men in 1615, when Robert Bylot and William Baffin, who was Bylot's mate and associate, voyaged in the region. Baffin gave the name *Seahorse* to the rocky promontory because of the great number of "morses" [1] that the crew had seen on the ice near by.

Seahorse Point is rough country. Some of the cliffs are sheer rock-faces two or three hundred feet high. The lakes are small and deep. Fjords indent the coast. The currents of Fox Channel to the east bring down from the north great masses of ice that lodge along the shore, even, sometimes, in midsummer. Off Seahorse are some small islands that you will not find on most of the modern charts. But these islands are clearly marked on Baffin's chart that was published three hundred years ago!

We headed for Seahorse Point on the day after we located John's whale and got the polar bear. John wakened me eagerly that morning, saying in a hushed voice, as he pointed upward: *"Toolooghak sittamoot tidlimoot,* Doctor!" [2] The birds had gathered round the bear carcass, hungry as usual. John wanted me to shoot one of them so he could watch it fall and see it hit the ground! More heresy in the offing, no doubt!

[1] Walrus.
[2] "There are four or five ravens out there, Doctor!"

The country became higher and more rugged as we moved eastward, and the water plainly deeper. In the far distance now rose the smooth dome of Mount Minto, not a high mountain at all, but of a dizzy height as compared with the flat country at Cape Low. Snow was falling. In mid-afternoon, as we were rounding a rocky sort of cape that wasn't indicated on our chart, we sighted two bears, one of them on a columnar cliff not far from us. We stopped the *Shookak* in a twinkling, let down the anchor and put to shore in the tender. (Jack and I by this time were calling the tender the *Peanut Shell.*) We made brief plans as we rowed shoreward. John and Kayakjuak were to walk inland and try sneaking up on the bear. The rest of us were to wander about in the open, in plain sight all the time, so as to keep the bear's attention focussed on us.

Whether the bear ever saw us or not is a question, for I am certain a polar bear's eyesight is not especially good. But the plan worked. Half an hour passed and the bear was dead, his massive body heaped into a fissure in the rocks where he had fallen. It wasn't easy lifting him out, for he must have weighed three or four hundred pounds, perhaps more. He had been eating berries, for his jowls were stained a sort of orange-red. The Eskimos skinned him rapidly, and we carried some of the red meat back with us. As we walked to the tender we passed over some strange subterranean caverns. We could hear odd echoings under us when we stamped our feet.

"Queer island, this Shugliak!" I found myself thinking. And the Eskimos assured me there were many such "hollow places" all along the eastern shore.

We reached Seahorse at evening. A great flock of *nowyah* [1] circled round us. Snow covered the granite cliffs and hills

[1] Gulls.

like a film of gauze. We climbed one of the high points near our sheltered anchorage and made out the forms of five bears on the cliffs. We had fried bear steak for supper. It was a glorious meal.

Next day we hunted bears. This country was full of *nanook.*[1] We got four. We saw eight at one time, eleven at another time, all in sight at one instant. No wonder Sam Ford had called Seahorse the "home of the polar bear."

NANOOK: THE POLAR BEAR

Polar bear-hunting at Seahorse Point isn't always as exciting as you may imagine it to be. You see your bear and walk toward it, wondering all the time why it doesn't run away. "Is the brute drunk, or asleep on its feet, or what is the matter?" you ask yourself, and then a rifle is raised, you hear a sharp *crack!* and the bear comes rolling down the rocks. Carrying back great chunks of meat and bundles of hide is gratifying within certain limits, but it becomes drudgery after a while. There ought to be more growling, you think; more running and chasing, a little more excitement. I continued to

[1] Polar Bears, *Thalarctos maritimus.*

like bear steak very much, especially the fatty part; but bear-hunting frankly palled on me a little.

By this time, of course, you are ashamed of me. You are thinking me a mere hound for danger and punishment. Well, you soon will see how the *tornyak* [1] dealt with me for finding fault with polar bear-hunting and its lack of thrills.

Retribution came within the next twenty-four hours. It was a cold, dull day. I was off by myself on a lonely promontory, the eastern side of which was a lofty cliff. Fog veiled the sea. I had been pursuing ravens and glaucous gulls and falcons and had had some success. I carried my twelve-gauge shot-gun.

As I made my way round a slippery ledge I noted to the northward a loose flock of gulls circling over a rocky mound half a mile away. A family of ravens joined the gulls as I watched. It was natural to wonder what should attract so many birds in a region where birds are scarce. I started for the mound at once.

I had crossed the thin ice of a narrow lake and made my way through a deep defile when I saw him, noble *Nanook*, golden-white in color, massive and dignified, pawing at the thin turf for mushrooms and dried berries. I tried to conceal myself so as to be able to watch him; but already he had sighted or scented me and was trundling up the rocks, his flat head turning backward now and then in my direction, the upper lip extended and nose lifted in an odd grimace. He disappeared behind the mound.

I followed, cautious but vastly interested. When finally I reached the top of the mound I looked down upon a long, narrow bay, the innermost fjord of which was packed almost solid with green and white ice. On a level with me a few

[1] Spirits.

gulls circled slowly. One was a handsome, pearl-mantled *nowyavik*,[1] an adult with yellow bill. The ravens, at sight of me, came close for a moment's inspection, then wheeled away, yawking angrily. I observed that what had attracted these birds was the half-eaten carcass of a *kiolik*-seal that lay among the rocks a hundred feet from the water's edge. Eager to examine this carcass (it was the first of this species I had seen) and forgetful, for the moment, of the bear, I clambered hastily down the ledges, trusting that one of the rarer gulls might come close enough for a shot.

My nape tingled! Near the seal, his body bottle-green in the sea water, his head turning slowly from side to side, was *Nanook*, and the *kiolik* was his! Scarcely had I halted in my descent of the ledges when he pulled himself from the water, shook himself once or twice to the great annoyance of the gulls, and looked toward his prey. Then he ambled up the rocks—not only toward the seal but directly toward me!

I was impelled to crouch, then to flee, trusting that my legs would carry me so fast that my pursuer would give up the chase before I should be winded. I forgot completely how docile and spiritless all of yesterday's bears had been. Would he follow slowly but tirelessly, as a weasel follows a mouse, to the death? Would be bound up the rocks once an eddy of the breeze carried knowledge of my exact whereabouts to him? My knees trembled; but I watched as if charmed. I could not take my eyes from him; he was so majestic, so imperturbable.

Suddenly I realized, with a blessed sense of relief, that the steady breeze was blowing from the bear toward me— not from me toward the bear. Moreover the bear evidently had not seen me. At once I was planning counter pursuit. My shot-gun was a good one and I had some shells loaded

---

[1] Glaucous Gull, *Larus hyperboreus.*

with Number 4 shot. Would these pellets penetrate a bear's hide or skull? If so, how close should I have to be? If the first shot failed to kill and only incited the maddened creature to attack me, could I steady myself for a second shot at close range?

"Perhaps you had better not tamper with that *Nanook!*" I thought. "Why not just watch him, and jot down some on-the-spot impressions in your note-book?"

"But how fine to be known all over Shugliak as the white man who killed a huge bear with only a shot-gun!" Vanity, all is vanity; even at the Arctic Circle!

Slowly I began to stalk my prey. *Nanook* had by this time reached the seal. He stood over the headless body contemplating his repast. Foot by foot, inch by inch, I crept from ledge to ledge, closer and closer.

How large was this bear, anyway? How close was I now? A hundred feet?

A gull dipped down, trailing its feet languidly only an inch or two above *Nanook's* black nose. The bear half rose on his hind legs, made a pass at the gull, then sank to all fours. Gulls continued to pass, hinting for a morsel of food, and *Nanook* was plainly annoyed at each and all of them.

Down, down, nearer, nearer, over moss and lichen—sliding, scraping, inching along—to my relief I perceived that if the bear should attack me he would have to scale a perpendicular wall of rock at least fifteen feet high. How strong and agile was he? How much did he weigh?

He turned his face toward me. Quaking with eagerness, exaltation, or fright, I "froze" in my tracks. I scarcely breathed until he turned away from me to gaze out at the bay. More gulls dipped down. As his eyes followed their lazy flight I jerked forward rapidly. Now I could go no farther without descending the fifteen-foot wall of rock.

Across this little chasm, probably not more than sixty feet away, was *Nanook*. He looked directly at me, but his face was mild. I began to wonder whether bears ever see anything. The situation was a little comical. I was very cold.

Nanook turned a little, grunted once or twice, and put one of his front feet oddly on the other. He sniffed at the seal, lifted his nose to the gulls, looked blankly at me, and sniffed the air again.

I heard a distant rifle shot. *Nanook* turned. There was another shot, a second elapsed, then came the sullen *plop!* of a bullet in the water. *Nanook* started. The Eskimos on the other side of the bay were shooting at *my* bear: at this bear that I had been stalking for three-quarters of an hour! I raised the shot-gun deliberately and fired the left, the choke, barrel, at the massive face.

Nanook shook his head like a fiend and rose on his hind legs with a horrible hollow growl that issued from a cavernous stomach. Blindly he struck with his sledge-hammer paws, swinging round dizzily, roaring like a saw-mill. Blood streamed from his face. He growled again, then—horror of horrors!—took a short bound in my direction. He was scarcely even stunned. My shot-gun was only a toy. He turned and made his way, a little haltingly, up toward the higher ledges to my left.

Teeth chattering, I said aloud, but in a low voice, "Well, it's your turn!"

I don't know exactly what I meant by this statement. But I said it, out loud. I was prepared to take what was coming to me. I was completely ashamed of my ignorance concerning bears and fire-arms and ammunition. Why hadn't I had my rifle? I was sorry I hadn't killed the bear, but I was somehow far more sorry that I had hurt that handsome animal because I wanted to be famous for shooting a bear with a shot-gun. I was a little afraid, too,—afraid the bear would run only far enough to get my scent on the wind, then turn to come for me, this time much to my disadvantage.

I tried to rise. My right leg was cramped. But I made my way up the mound, gazed gratefully at the retreating *Nanook,* and sat down for a rest. I had had quite enough bear-hunting for one day.

Half an hour later, when my legs had limbered and my pulse slowed down to normal, I followed the trail of blood and was glad to learn that the animal was not bleeding badly. I walked to the opposite side of the promontory where, in the damp sand, broad tracks led to the sea. Out in the gray waves *Nanook* was swimming for a distant island, his head low in the water, black furrows edged with wavelets following the glistening whiteness of his neck. I could barely see him through the thin fog.

I started my long walk back to the *Shookak.* As I reached the edge of the sand a lithe weasel, his brown coat blotched with the white of winter, darted under a rock. Attracted by the drops of blood, he too—fearless, headstrong, redoubtable butcher that he was—had been following the wounded *Nanook* toward the sea!

Chapter X

Devil's Gorge

My SEALSKIN *komik* were wearing out. This rough country at Seahorse, these excursions after bears along the falcon cliffs, were hard on foot-gear. I walked a good deal more than the Eskimos. Eskimos do not like to walk very much. They prefer to ride on *komatik* or motor-boats. And they like exceedingly well to be carried across shallow rivers on some tall white man's back.

We began our journey "home" from Seahorse on September 24. It was becoming colder. More snow had fallen during the night, enough to give a piebald appearance to the landscape. I had managed somehow to make a sketch-map of the region. It was not a very neat map, for my hands were cold when I made it.

We made our way straight to John's whale, there to take on a load of *shookak*.[1] The Eskimos set to work with knives and hatchets, trimming off the fringes and peeling down the long, flexible strips, laughing and talking as they worked, going over and over the details of the shooting of their various bears at Seahorse.

I went for a walk, as usual, flapping along in my patched *komik,* hoping to find some new birds. I was much interested in the appearance of the cove at low tide. Here the rock bottom had the appearance of a vast mosaic of lateral cross-sections of gigantic cabbage heads, the thick rock-leaves

---

[1] Whalebone, baleen.

arranged in concentric, symmetrical circles. These rock-leaves were exceedingly difficult to walk upon, partly because of the mud and the kelp, partly because the broken surfaces were tilted at so many different angles.

All over this exposed flat, with its strange cabbage-heads and narrow, curved pools of water, shore-birds swarmed: sanderlings, turnstones, peeps and plover. When I returned along the beach at evening, I would startle sleepy little

AKVIK: THE GREENLAND WHALE

bands of them that had gone to roost in the shelter of low boulders or among the kelp beds.

We had a good load of prepared baleen now, as much as the motor-boat could carry. The untrimmed whalebone we shocked on the beach. Since the wind was rising and a storm apparently brewing, we moved into the mouth of a fair-sized stream near by, anchoring in the shelter of the western bank.

We sighted a giant whale far out in the bay. At first the great beast merely spouted high jets of water that had the appearance of geysers. Then he began leaping. Fully twenty times he must have leaped, sometimes clearing the water completely, like a vast black submarine gone mad. When he fell back the splashing (which we could not hear) appeared to drench the very clouds.

When I spoke to the Eskimos about this peculiar behavior of the whale, a long discussion began. I couldn't follow the various nuances of this discussion, of course; but Jack explained them to me in part. They had to do with the great Sedla, Goddess of the Sea, a deity I longed to know more about; with the sayings and doings of the *angekok*, the soothsayers, of the Aivilik tribe, and with the beliefs of the "Old People," the Shugliamiut, who had been primarily a whale-hunting race.

"Why do whales leap out of the water the way they do?" I asked, requesting that Jack interpret for me. "Is it because they are mating, or playing, or are they being attacked by killers?" I had the expression of the seeker-for-truth on my face.

"It is mostly because they are constipated," John answered. We all laughed. And I suppose I shall never know whether John was serious or not. I somehow cannot think of whales as sufferers from any form of indigestion, but then I don't really know anything about whales, and I can readily believe that calisthenics and setting-up exercises are as good for whales as they are for anyone else.

High wind forced us to remain at this river-mouth four days. The Eskimos spent a good deal of time telling stories and sleeping for there was nothing much more exciting than this for them to do. I should have enjoyed staying with them, just to learn more about them, but I was busy. This would probably be my only chance to see the southeastern coast of the island. So I walked up and down the beach for miles, trying to learn more about these shore-birds that were flocking everywhere; these brant that were moving westward; these duck hawks that sat about on the boulders waiting for their crops to become empty enough to permit them to dash after and cut down another sandpiper. Voracious killers, these *kigaviatsuk!* [1]

I decided to go inland. The farther I progressed from the coast the more barren I found the limestone plateau. Here were mile upon mile of gray-white, angular chunks of rock, with scarcely a blade of grass or bit of moss anywhere. No willows for the lemmings; no lichens for the hares; no berries for the ptarmigan; nothing anywhere but these rough chunks that you stubbed your toes upon and that slipped and rattled and turned under your feet! It is easy to find yourself just walking along in such a place, moving straight across the barrens, not thinking about anything in particular, a little tired of it all, a little disgusted at yourself for making your way across a continent just to reach a place like this.

A dull sound made me stop. I had been hearing this sound now for quite a time; but it was an indefinite sound, a sort of low roar that had been so faint I had not at first really been aware of it. It was louder now. It was a strange roar, hollow, muffled, more distinct with the gusts of wind.

A waterfall? A waterfall in this barren place?

I walked forward a quarter of a mile. The roar became louder. It sounded, now, almost under me, off an indefinite distance through the ground. There was a dry rattle and a crackling and a dull *boom!* Another rattle, another crackling, another *boom!* louder this time. What was going on in this weird place?

I went forward cautiously. I became aware of a wide rift ahead of me. Now a chasm yawned at my feet. Seventy or eighty feet below me, at the bottom of this chasm, spread out a broad, deep pool, green and clear as emerald. Into this

[1] Duck Hawks, *Falco peregrinus anatum.* The American race of the famous Peregrine Falcon of Europe.

plunged two white falls, each of them issuing from a narrow, lofty gorge. Once more there was a rattle, a sort of pitter-patter of loose gravel, and a sullen *chug*. A section of the chasm's rotten wall had given way, falling into this emerald pool. There was an almost continuous sound of falling chunks, those farther downstream striking ice instead of water. Instinctively I stepped back; then, fascinated by the memory of that clear greenness that was so breath-taking in this colorless country, I returned to the edge, peered down into the depths, tried to see the bottom. I moved a boulder carefully, not at all eager to jar this crumbly place unnecessarily, and let it fall over the brink. It struck the water, shot upstream as it was caught in a current, moved over to one side, then to the other as it made its way down, down, down, and then, still going downward, faded out of sight. Was this just an extraordinarily deep pot-hole, or was part of the river actually disappearing underground?

I walked downstream, keeping at what I considered a safe distance back from the edge. A narrow-winged falcon beat rapidly past me, showing the blue-gray of its back as it shot down into the chasm and disappeared. Here the calm pool was frozen over, the ice patterned with symmetrical whorls, especially along the opposite side. These whorls were strange and beautiful, but they appeared treacherous. Rocks continued to fall, making booming sounds. Occasionally boulders that broke off were large enough to crack the ice and other, sharper notes were added to the wild symphony. I can't say I exactly enjoyed this gorge; but it was a thrilling place.

"Jack must see it!" I said to myself.

From the ice at the lower end of the chasm the river boiled out, making its way through a narrow bed along the very base of the opposite cliff. I made a detour, clambered down over the rocks, and returned to the stream at a considerably lower level. Inching cautiously along I walked forward. The ice was clear, startlingly clear, and perhaps a foot thick under me. Darting this way and that beneath the ice were a hundred little gray trout. How oddly cozy and safe these slim gray beings: safe from every enemy, I suppose, save themselves! After all, what *could* they find to eat in this barren place save themselves? And that being the case, how did this race of trout ever start or ever maintain itself? My train of thought stopped right there, for want of facts.

I was in the deepest part of the gorge now, looking up at the precipitous walls, thankful I did not have to stand directly under them, under those falling boulders. The falcon flew out. I saw the white-washed place on the rocks where it had been roosting. On the ice under this white-washed spot were the remains of birds it had been eating—principally the wings and breastbones of shore-birds.

When I got back to the motor-boat I was full of news. Jack had never seen the gorge and was eager to make a tour of exploration. But the Eskimos had seen it and they said the river was a raging torrent in spring, so deep and big that the falls and gorge were completely submerged.

Jack and I went back next day to fish. We chopped holes in the ice; put down all sorts of dainty morsels and bits of bright cloth and shining objects on little hooks; and stood very quietly on the ice so as not to frighten the little creatures. But those little trout would not be caught. I wonder, even as I write, if they may not be of a distinct species of Arctic charr, indigenous to Shugliak. It will probably be a long time before any one knows, for I did not get a specimen.

We decided to call the gorge Devil's Gorge. We somehow couldn't help calling it that, though we realized there was nothing very original about such a name, and felt that the devil was getting more honor by way of Geography than was really due him.

On the 27th we left Devil's Gorge, headed for the post. The front of the boat was piled with whalebone. Near Kik-kuktowyak Island we got a dovekie, a rare bird the Eskimos called *akpaliatsuk* or *akpaliatclook*,[1] a wanderer blown in from the Atlantic by storm. At Native Point we said good-bye to Kooshooak and looked at two *oogjook*-seals the Es-kimos had got.

On the way across Native Point Bay we ran into a big drove of walrus and there was more of this feverish *aiviuk*-chasing. I was sorry it was so late in the day. It would have been great sport to watch the brutes a while longer.

We reached the post on the following day. We met Sam and Angoti Marik out in the inlet hunting *netchek*.

There was a long talk about Seahorse and Devil's Gorge that evening. All the Eskimos at the post came in. We learned that three *kellilughak* and a *netchek* had been caught in one of the whale nets at Seal Point.

Winter was at hand. Most of the snow buntings were gone. A few purple sandpipers still frequented the half-frozen piles of seaweed, and small flocks of pipits were to be seen in the marshy spots. It was no longer *Ookiukshak*, early Autumn; it was *Ookiak*, the "time when the islands in the bays freeze shut."

---

[1] Dovekie or Little Auk, *Alle alle.*

An iceberg off Southampton Island.

Aivilikmiut Eskimos at Coral Inlet.

John Ell, whose Eskimo name is Amaulik Audlanat, with a huge polar bear skin.

Old Shoo Fly, matriarch of the Southampton Island Aiviliks.

By and By, an angekok of the Aivilik Eskimos.

John Bull.

An Aivilik Eskimo girl.

A husky puppy.

Captive polar bear cubs.

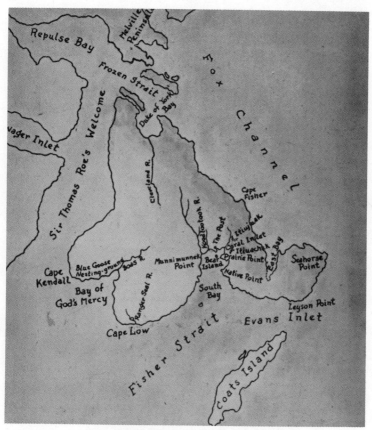

Southampton Island map.

A mother Snow Bunting on her nest.

A young Snow Bunting.

An Arctic Tern home from the Antarctic.

Netchek or Ringed Seal stranded on ice.

A Long-tailed Jaeger on her nest.

A Semipalmated Sandpiper.

Chapter XI

Ookiak, the Late Fall Season

THE Aivilik Eskimos have a colorful language. In their poems, in their tales, even in their casual conversation you may note figures of speech that evince a vivid imagination. But when you hear the Aiviliks talking about the month of October you hear nothing of "bright blue weather." You hear of *Nooliakvik*, the season when the bull caribou scrape the velvet from their antlers and battle for supremacy and mate. You hear of long *komatik*-trips across the frozen lakes to the highlands where these caribou live. You hear a clanking of rusty fox-traps that are being dug out of stone-bins in the *tupek* and pulled from dark corners in the servants' house at the post. And you look out to see a film of leather-ice stretching across the inlet.

At the trading post, unless the weather is stormy, there is always a bustling on the first day of October. Every available dog is hitched to the heavy, floating wharf down near the oil-shed; all the Eskimo servants, the women included, give themselves over to a grunting and a puffing and a sweating; and the wharf scrapes and bumps and crunches its way up to its wintering-ground near the store. Unless this work has taken too much time and energy, everybody gets at the motor-boats and they too are hauled up to some safe place on the shore. Storm-windows are nailed on the dwelling houses, and some of the windows of the store are boarded-up. Big barrels that will hold the winter supply of ice for drinking-water are rolled into place. If the whale-nets are not catching many *kellilughak* or *netchek*, these are brought in, spread out for drying, and folded up. It is not an easy task, this pulling up and bringing in of the wet, cold, stiff, heavy whale-nets.

OOKALIK: THE ARCTIC HARE

Jack and I decided we'd go hare hunting. I hadn't yet shot an adult Arctic hare and word went round that Sam was an expert at baking pie. We walked a long way that day, and saw little of interest save the remains of a partly white ptarmigan that probably had been killed by a snowy owl; a duck egg that had been buried, then dug up, by a fox; and a few longspurs and pipits. We were walking across a long lake when we noticed a little patch of white on an exposed rockledge on the shore opposite. Jack said it might be a hare, but that he could never really be sure about such things; and I said I didn't see how it could be a hare, right there in plain sight that way, but that I couldn't really be sure about such things; and so we walked over to it, curious as weasels, laughing at ourselves for being so gullible, and making little statements by way of explaining that we were not really being fooled. Eventually we came within stone's throw of the white patch that certainly did not in the least look like a hare.

"It's just a chunk of snow," I said; and even as I spoke we noted the gentle waving, in the wind, of deep, downy fur. In an instant the hare, huge black-tipped ears lifted straight, was bounding away, its great hind feet pushing it powerfully upward and forward. Both rifles cracked, but the hare kept going.

Jack, knowing the way of hares, ran back along the edge of the lake. I followed the hare but caught no more than a fleeting glimpse of the creature bounding lightly along the edge of the ice. When I reached a high point among the rocks I looked out and perceived at once that the hare was not crossing the lake. Then I heard a shot. True to form, the animal had circled back among the rocks and Jack, waiting for it behind a little ridge, had got it without difficulty. It was a beautiful creature, in complete winter pelage, about the size of a fox terrier. Its eyes were deep brown.

You mustn't think it always a simple matter shooting an *ookalik*.[1] Sometimes the hare goes straight across the lake and on and on for miles before it begins to circle back. And sometimes it runs out onto the frozen bay and makes its way round the ice pinnacles, leading you the merriest of chases. But Jack somehow sensed that this hare would make its way back after a short run forward. And Sam's pie was grand. We had it for supper, then again for luncheon at midnight, and again for breakfast the next day.

·   ·   ·   ·   ·

At breakfast there was considerable talk about Ginger.

[1] Arctic Hare, *Lepus arcticus*.

Ginger was a dirty-orange dog that lived at the post, rather an old dog, not very prepossessing in appearance, and, according to Sam's latest dictum, absolutely worthless.

We decided to kill Ginger with a dose of arsenic on a piece of seal-meat. This method probably was not as humane as it might have been. Maybe we should have decided to shoot him. But for one reason or another we didn't. Sam got the meat and I got the arsenic (about three tablespoonfuls of it) and we gave Ginger his "medicine." He gulped the chunk of meat down, perhaps wondering dimly why he should thus be proffered a special meal. The incident was not without its complications, either; six or eight dogs fought about that piece of meat long after Ginger had swallowed it.

We weren't exactly proud of the way we watched the old dog, seeing if he would die calmly or in convulsions. He went to the side of the house, calmly lay down, and went to sleep. His breathing appeared to be normal. We more or less forgot about him until about two hours later when we saw him vomiting near the water-barrel. We went out to watch him the more closely and he ran nimbly away, apparently afraid he was going to be given another special meal. He didn't show up for about two days. When he appeared again he was a new animal, vivacious and chip-on-the-shoulder, as a good Husky ought to be. He was still flop-eared and dirty-orange, but in other ways we hardly recognized him. So this was the worthless Ginger—poor dog, too bad to have to kill him. Too bad, indeed! I think everybody at the post wondered if I didn't have Sugar of Youth or something of the sort in that can marked with red skull and cross-bones.

. . . . .

On the fourth day of October, John Ell and Kayakjuak started on a caribou-hunt. They took John's big *komatik* and most of the dogs. I wanted to go along, but feared I might not be much of a help, and furthermore was very busy in collecting specimens in the vicinity of the inlet.

. . . . .

On the Saturday night following, I received many messages over the radio. Station KDKA, in Pittsburgh, with lively, rapid-fire Louis Kaufman and his fellow announcers at the microphone, sent words of greeting from home and friends back in the States and in southern Canada. I was not lonely, not really in need of messages; but it was pleasant to hear those cheering words, to know that my family were well, to realize that I had not been forgotten. And it suddenly dawned on me that all this talking over the air was making the eyes of the Eskimos fairly pop out of their heads. Sam was good enough to translate for them so they knew what my world was saying. After twelve or fifteen messages had expressed the hope that I was having success in getting all the birds I wanted, the Eskimos plainly began to feel that their decision to help me get specimens was justified; that this "Doctor" who had so mysteriously come among them must be a great person indeed if the spirits were willing to carry all these messages to him.

The Eskimo probably does not understand the White Man's radio at all. He is not worried by it, now that he has learned it is harmless. Whether he still really believes that spirits are carrying the messages or that tiny people are producing the sounds is more than I can say. But he has a certain respect for phenomena of this sort that appear to have a spiritual significance. The radio is taken for granted when it gives forth music as does a gramophone; it is more impressive when it says something about the doings of the Outside World in such a manner as to hold Sam Ford's undivided attention for a time; and it is indeed very impressive when it brings special messages to "the Doctor"; many, many messages; a whole hour or so of messages; messages that make Sam and Jack and "the Doctor" all listen seriously and hold their hands up for silence when anyone makes a noise. The Eskimos could tell from the expression on our faces that we were bound to another world while those messages were coming through. There could be no faking such an interest as this.

* * * * *

I was skinning redpoll specimens the following Monday when I heard that clamor and bustle that meant a white gyrfalcon was in sight. I heard Jack shouting from somewhere, "A white hawk, Doc! The people sees a white hawk out there!" And out we all rushed, knocking over chairs and boxes as we grabbed our shot-guns.

The falcon was there, all right, trying to alight on the flagpole up near the Mission. I ran forward. Jack set up a loud outcry behind me, cackling like a ptarmigan. Soon the chorus was taken up by all the Eskimo children and such a gobbling of ptarmigan you never heard. The gyrfalcon left his lofty perch and came straight for us. I shot, and the handsome creature tumbled with a broken wing. I ran up not a second too soon. The dogs almost beat me. I managed somehow to get my bird without letting it slash me with its sharp talons, and then, as usual, the disappointed dogs began to fight. They dared not nip me for taking that bird away from them, but they could nip one another to their hearts' content. The gyrfalcon was a lovely specimen, the plumage of its back barred and marbled with ashen gray.

* * * * *

Back of the Mission building, on a little hill, was the Catholic cemetery. Here the piles of rocks that covered the bones of the faithful were marked with thin, black, wooden crosses. I was passing the cemetery one day when I saw a weasel. It was not yet fully white. Its den apparently was in one of the graves.

. . . . .

Some of the Okomiut Eskimos came in from Cape Low bringing several bear-skins; but these skins were worthless because the dogs had chewed off the noses and ears. The Okomiut were careless about such matters. Sam would tell them over and over again that mutilated bear-skins were worthless for "trade" but they would bring them in saying they had nothing to eat and Sam would have to tell them again. Being fatalists they apparently felt that if the dogs chewed up and ruined their bear-skins it was a matter for the spirits to rectify if the spirits cared to; and if the spirits didn't care to rectify the matter, complaining would not help. "*Kooyannah ayornamut,*" they would say. "It doesn't matter; it can't be helped." And this philosophy they voiced over almost any accident that might befall: the cutting of a finger, the breaking of a *komatik*-runner, the drowning of a wife or husband.

. . . . .

In mid-October Sam built an ice-boat, the first in the history of old Shugliak. From my work table I could hear the sawing and the pounding, and the jabber of Eskimo women downstairs telling one another how the canvas sail should be sewn. Some of the big lakes were frozen so smoothly and the ice was so free of snow that Sam thought a little ice-boating would be good fun. Father Fafard had been skating many times already.

Sam worked a long time at that ice-boat. Although I am no authority on the construction of such craft, it was a good-looking ice-boat to me. Furthermore, Sam sailed it, and Jack sailed it, and I sailed it—a little. But one of the runners was weak or the bar controlling the rudder was not quite sturdy enough: at any rate, Shugliak's ice-boating season for 1929 was a short one. Much snow fell during the next few days. In a week the ice-boat, which had been pulled into a little cove, was buried, all but the top of the mast.

. . . . .

The displays of aurora borealis were becoming more brilliant and more frequent. Almost every night now the mysterious lights flared and quavered above us, shifting this way and that, waving like weighted curtains or shooting out like sheets of luminous snow blown across a vast frozen lake. Frequently the display had the appearance of slow-moving clouds. Usually the lights were very pale yellow with a delicate pink or purple or green band along the lower margin.

The Eskimos at the post paid little attention to the aurora, apparently making no attempt to interpret its various moods in foretelling the weather. They told us tales a plenty about the spirits that inhabit and control the lights, however, assuring us that when the display is brilliant a game is being played, a game in which a ball is kicked and tossed about through the sky. And the ball is the skull of an Eskimo not long dead. They told us too that if we were to whistle to the lights they would come closer.

"Jack, let's go out and whistle to the lights, just to see what will happen," I suggested.

. . . . .

So we went out to the flag-pole and stood there hissing and whistling and cat-calling for about fifteen minutes in the crisp night air. We made quite a noise. Had we been standing on a sidewalk in New York a taxi would have dashed to the curb and stopped with a screeching of brakes, or policemen would have come on the run, or ladies would have turned to look, or dogs for blocks round would have pricked up their ears. But out under our flag-pole nothing happened. No dog paid the slightest attention. The plaintive sound of the little organ at the Mission continued uninterrupted. There was not the slightest dimming of the steady light from the window at the house. And the aurora made no gesture of recognition. A perfectly plain case, of course, of justice to the faithless. Had we been believers, had we been good Innuit, the aurora would have reached downward toward us, perhaps with a faint sound of hissing and crackling.

•   ■   •   •   •

On October 16, Angoti Marik came back from a lake called Salmon Pond. He brought with him several burlap sacks full of two-foot *ichalook-* and *eeveetahgook-*trout, the former slim and silvery with white or pinkish flesh, the latter broad and scarlet and blue-spotted, with protruding, hooked lower jaws that made their faces look like a bulldog's. These fish were so different in appearance it seemed scarcely possible they could be of the same species.[1] The *ichalook* were delicious fried. But the *eeveetahgook*, handsome as they were, were fit only for dog-food.

•   •   •   •   •

On the following day great flocks of migratory ptarmigan arrived, all of them in full winter plumage, white with a

---

[1] The slim, silvery *ichalook* were female; the scarlet, hooked-jawed *eeveetahgook* male, Greenland Charr, *Salvelinus stagnalis.*

lovely shell-pink cast, save for the black tail feathers that never showed until the birds flew up and away. There were two species of these northern grouse, rock ptarmigan and willow ptarmigan. The Eskimos called both species *ahigivik*, though they had also a separate name for the "rocker," that called attention to its "belching" cry. The *ahigivik* were very tame. They sometimes came into the post in great numbers, running this way and that as if they had never seen a house, or a dog, or a man. Indeed they may never have seen any of these, for their nesting ground was the unknown wilderness of northern Shugliak, of Melville Peninsula still farther north, or of the islands north of that. They were definitely migratory in the region of the post, not many of them summering there, for the Eskimos killed them constantly. During some years, so Sam Ford told me, thousands of them had appeared in late fall or winter, living wherever they could find enough willows.

. . . . .

We ate ptarmigan frequently. They were rarely fat; the flesh sometimes had a slightly disagreeable flavor.

. . . . .

Keetlapik, the Eskimo convert in charge of the Anglican Mission, sent me a box full of *netchek*-heads, the bones badly broken, most of the skin gone, but the whiskers, the eyes and the tongues all there. It was an astonishing mess, that pile of seal-heads. The skulls, with their great, round eyes, made me think of the head-on aspect of some strange foreign make of automobile. As specimens, they were worthless. But I was glad for Keetlapik's interest, so I kept them, bearing in mind that I could now start a dog-fight whenever I felt the need of extra entertainment.

Late in the evening, on October 19, John Ell and Kayak-juak came back from their caribou-hunt. They had got one caribou and two bears at East Bay. They hadn't seen many birds, only an owl and one or two ptarmigan. They told us Muckik had got two bears.

I was having trouble these days with a sort of rash that broke out on my face and neck, especially in the region where the coat-collar rubbed, and where my cap came down across my forehead. I felt well, certainly was getting plenty of exercise, and did not suffer much with the rash; but since I had never had such an eruption of the skin I worried a little about it. I fancy the condition developed as a result of same-ness of diet. Sam told me that everyone wintering in the North Country for the first time had troubles of the sort. I applied various salves, washed carefully, and tried a change of food. Sometimes I wondered if I were in the incipient stage of some dread malady.

．　　．　　．　　．　　．

We had some wonderful discussions of diseases. Sam told me of a white man in the Far North who had suffered from a frightful pain in his right side and who had eventually frozen this area and cut out his own appendix. He told me, too, of an Eskimo who had gouged out a decayed tooth with a piece of iron from a barrel hoop. We talked about scurvy and pneumonia and influenza and syphilis, sparing no de-tails. Sam considered the Eskimos, on the whole, rather a healthy lot. He had had few illnesses of any sort, himself, while in the North, though he occasionally had suffered from a sort of rheumatism and had made one trip to the Outside to have his appendix removed.

．　　．　　．　　．　　．

We talked about old Shoo Fly, John Ell's mother. On the day I first saw Shoo Fly I had noticed a frightful sore on her neck: a deep, pale-edged ulcer that was partly covered with gauze, and that she frequently pressed with a wad of cloth held in her hand. This sore sometimes healed, Sam said, but it would break out again, usually on her neck or chest, and she would have to bandage and care for it as best she could. This was the only such sore I saw on any of the Eskimos of the island. Kooshooak's sister had a deep-red area on her jaw that had somewhat the appearance of a cancer, but this area had not caused any trouble.

One of the women brought me three floppy, greasy new-born-puppy skins. I hastened to make it clear that I wanted no dogs. I had unpleasant visions of Shugliak turned into a vast dog-ranch, everyone urging the dogs on to mass production so as to keep "the Doctor" amply supplied with specimens.

On October 22 I took my first long *komatik* trip. John Ell and I went to the head of the inlet on a seal hunt. It was a glorious experience for me, watching the dogs, observing John's driving, and listening to the strange words and sounds he used. We were gone a day.

When you drive dogs on Shugliak you frequently say something like a long-drawn-out "Aha!": almost exactly the sound you make when you catch one of the children with his hand in the cooky jar and his mouth all ringed with crumbs after you have been wondering for days where the cookies were going. Then you say something like "Howk! Howk!" which is supposed to indicate your desire to go to the right; and an "Ouka! Ouka!" that means "To the left!"; only when you pronounce "Ouka!" you leave the *k* out, making it two, explosive syllables. Then you cough and mutter and growl, not attempting to express any special command to the dogs but merely, I should say, letting them know you are alive and well and that they are not getting away with anything. You must keep up a running conversation. If you don't, if the dogs begin turning round, if they begin letting their traces drag, they eventually stop.

Then of course you have to crack that twenty-foot walrus-hide whip, making a fierce and significant threat with every crack and upon occasion taking out neat tufts of hair by way of emphasis.

John and I conversed whenever we went through country where the snow was smooth and where driving was therefore not difficult. He knew a little English, you see, and I knew a little Eskimo. He was fond of using such expressions as "Too many wind, Doctor!" Once, after three of the dogs had paused to regurgitate some seal meat he said, "She too many belly full!" Smile at John's English if you will; but reserve your laughter for the time when, having yourself learned to speak Eskimo, you hear John tell of what I tried to say to him.

When we got to the head of the inlet we looked out over the ice and water for seals. We saw six, all of them quite a distance out. They appeared black, but John said they all were *netchek*. He went after them while I walked inland. I turned now and then to watch. John was leaping across cracks, making his way forward behind chunks of ice, stopping whenever a seal lifted its head between naps.

I followed what appeared to be a stream bed that led into a vague white world. I saw blue-shadowed tracks of hares and weasels and foxes. Now and then I stepped into a willow, its topmost twigs just showing above the snow. What a country —with willow-twigs for landmarks!

You walk through whiteness of this sort for an hour or two and your eyes become weary. Since there are no trees, no telegraph poles, no water tanks anywhere, your surroundings do not step away from you as they should. It is a landscape without perspective. Here is the foreground, with its willow-twigs, its rows of fox tracks, and its bits of lichen that stick above the snow; and back of this hangs the pale tapestry of tundra and Arctic sky, a tapestry so strangely and so sharply marked with the tiny silhouettes of distant dogs and *komatik* and seals and boulder-tops and men.

You prefer not to think as you walk along, for thinking may lead you into an unpleasant frame of mind. "What are those black, shiny objects?" you half ask yourself, and you can't keep from thinking of medium-sized beetles wandering round, somehow, in mid-air; not flying exactly, but somehow moving. "Beetles? How could black beetles get into this picture? They must not really be beetles." And then it dawns on you that you are looking straight into the faces of half a dozen ptarmigan six feet away from you in the snow. You see them one instant; you don't see them the next. But you keep on seeing those black beetles that are their eyes.

And then you find the willow bushes they have been feeding on, the buds and terminal twigs all neatly nipped off. And you see the laciness of their footprints and the cozy beds they scratched out for themselves last night in the snow: neat little basins about six inches deep, all close together, and each with a neat pile of frozen droppings in the bottom.

I found some fox burrows along the edge of the snow-filled stream bed. There were several entrances to what appeared to be a considerable underground den. But I saw no fox.

When I got back to the *komatik*, that had been turned over so the dogs would not run off with it, I found that John had shot two seals. We fastened them to the sledge with *oogjook*-skin lines and started back. While making our way across the inlet we encountered soft ice across which we had to move in a hurry; and open channels across which the dogs had to jump, into which they fell, and from which they had to be pulled by their harnesses. It was exciting travel, this shooting across smooth ice, this bumping and whacking through rough ice, this leaping across cracks and hanging onto the sledge while it shot over deep fissures. I decided I had best not think myself an Eskimo just yet. I was immensely interested in being one, but I could not hope, I clearly perceived, to become one overnight.

Back at the post I was full of talk about our eventful day. We had roast caribou meat for supper and I ate a frozen slab of it raw: my first real *tooktoo-quak*.[1] It puzzled me a little, this taste of raw flesh in my mouth, this chewing down through frozen muscle fibres. I found myself, just for a moment, thinking of tapeworms. And then I fell to and enjoyed it heartily. This being an Eskimo was great sport indeed.

After supper I spent an hour trying to make the long whip crack. I made several different sorts of sounds that were not *cracks*, got myself wound up in walrus-hide a good many times, and dug one little hole in my cheek with the frisky, wilful, farthermost tip that had a way of striking any spot within a radius of twenty-three feet save the spot I was aiming at.

[1] Caribou meat; *quak* is almost a slang word.

Chapter XII

A Drowsy Weasel

You remember the granite ridge beyond the bog-cotton marsh just east of the post?

There was a big boulder near the middle of this ridge that stood high enough above its fellows to relieve the monotony of the skyline. On the sheltered side of this boulder the wind eddied so strongly that snow never buried its base; always between the granite and the clean-edged drift there was a narrow snow-chasm, smoothly curved, almost, as the inside surface of a clam shell. Here, when fresh snow had covered all the blemishes of the old crust, were frequently to be seen the neat tracks of a weasel. These were always in pairs, indicating that the animal never walked or trotted, but bounded along. The trail meandered—to no purpose, so far as I could see— all over the drift, here and there among the exposed rocks, and finally always to this highest point where the slender hunter probably surveyed his domain.

I was obsessed with a desire to see this *teggeuk*.[1] Did he hunt only by night? Was he as bloodthirsty as all weasels are

---

[1] Arctic Weasel, *Mustela arctica.*

reputed to be? Would he be in full winter coat by this time?
I decided to devote my attention to weasel tracks. I found
a trail and followed it for an hour. It led me hither and yon
along the ridge. Sometimes I lost it momentarily when it ran
into a snowless area or a large rock. Surely a weasel would become tired and hungry were he to keep up this purposeless
wandering all night long!

All at once I heard a startling sound, a sharp, staccato
chuckle like the barking of sparks in a physics laboratory. I
stopped, motionless. The trail of neat, paired tracks led under
a tilted, flat rock. Had I seen a flash of white in the blue
shadow?

I stood still and did the best I could to imitate the squeak
of a mouse.

I had not long to wait. Not only did the weasel appear; he
rushed toward me, looked at me with his glittering eyes,
flicked his black-tipped tail; and while I was marvelling at his
whiteness, his slenderness, his agility, and his wholly unexpected audacity, he as suddenly disappeared. I didn't know
whether or not he had retraced his steps. He simply was no
longer to be seen.

I squeaked again and instantly he reappeared. This time he
paused, lifted his long neck, stood on his hind legs, sniffed at
me, then flicked his tail and darted back. I began to realize
that the flicking of a weasel's tail is an important gesture.
While my attention was fixed upon that tail, the weasel himself was already *en route* to his place of refuge, and so agile
were his movements, so lightning-quick his leap to the
shadow of the rock, that I continued to watch the place where
the tail had flicked even though I knew perfectly well the
owner of that tail was already several feet away!

Then the weasel, as if asking me to put dull care aside,
came out to play! He mounted the rock; stared at the low

sun; took a look at me; sniffed the wind; bounced about gaily, his feet scarcely touching the withered moss; and somersaulted carelessly down the rock, biting at dry lichens as he went. One of these he caught in his mouth, crackling it with his dainty white teeth. Then he sat up and combed his tail briskly with his front feet.

More deliberately he bounded in high, graceful curves, to the rock's uppermost pinnacle; looked once more at the sun with slightly narrowed eyes; gave me one more glance; and yawned—right in my face! Then crouching a little, turning his head to one side in the manner of a mischievous puppy, he went limp, slid down the rock on his silken belly, and popped into his den. Though I squeaked like the most mortally wounded of mice, he did not come out again. He was probably sound asleep.

I might, that day, have recalled that the weasel, like the ptarmigan, carries "patches of snow" about on him while he is losing his brown summer pelage. I might have remembered that his winter whiteness makes it difficult for the fox, the snowy owl, and the gyrfalcon, all his enemies, to detect him; as difficult as for the lemming, the ptarmigan and other lowly creatures upon which he preys. I remembered not, however, the lithe creature's place in any biological scheme; but rather the sheen of his silken white coat, the trimness of his coral mouth and tiny teeth, his puppylike playfulness, and—that frank and impudent yawn.

CHAPTER XIII

South Bay Freezes Shut

DURING those latter days of October I was far more interested in becoming a good Eskimo than in achieving eminence as an ornithologist. I had, in fact, rather forgotten Ornithology, though I had no chance to forget bird-skinning. I grubbed away at the gull, duck and goose specimens I had shot, scraping off fat in a half-hearted manner (scraping off fat is not, at best, an inspiring occupation), and musing the while upon *komatik* and *tupek* and *tooktoo*-hunting.[1]

There weren't many land-birds, now that the most of the horned larks, longspurs, and buntings had flown south. Snow was gradually becoming deeper on the tundra. Wherever I walked, along ridge and meadow, countless small, round burrows led down through the drifts to the cool shadow-world of the lemmings. Everyone told me that it was to be a "mouse year"—a year when there would be lemmings *amishualueet*.[2]

. . . . .

A black bitch at the post gave birth to her first litter of puppies on October 24. She had pulled together a nest of excelsior and coal-sacks under the "office" window. Here her blind offspring squirmed and squealed incessantly, much to our annoyance. The bitch didn't understand motherhood, it

[1] Caribou-hunting.
[2] An unusually large number of lemmings.

appeared. She would cover her brood for a while, then rush off between the houses, nervous and afraid. The crying of the pups attracted all the dogs at the post. Finally, big Tweed, noting his opportunity, rushed in to the nest, snatched up a pup, and made off with it. We all heard the scuffle, noted that some of the squealing had stopped, and ran out to hurl stones. Upon being hit, Tweed dropped what remained of the pup with a yelp; but by this time Blackie, another of the dogs, was making off with the bloody morsel, and yet another dog was tearing in toward the nest after another pup. The Husky dog's early life, you perceive, is rather on the sad side.

We put the remaining two pups in the storm-porch while we made a new shelter, just outside the door, from a sledge, a ladder, and more coal-bags. The bitch covered her brood for a time but deserted them in the night. In the morning there were two small, icy carcasses for Ookpik to thaw out and skin when she had the time.

. . . . .

October 25 was Jack's birthday. Kayakjuak brought him a present, a combination ice-tester and harpoon made of wood and walrus-ivory, with a steel tip. This he could use in hunting *netchek*-seals in winter.

. . . . .

The night of October 25 was very cold. On the following morning the ice-sheet that covered the inlet was white and firm.

Jack and I took a *komatik* trip. Supposedly we were after seals. My personal feeling as I look back upon that memorable expedition is, however, that it was primarily a bit of schooling for young Bobbie and me.

Bobbie was a fine, year-old pup that had never been in har-

ness. Jack had decided it was time for Bobbie to haul his part
of the *komatik*-load, and that I was now enough of a North
Countryman to be able to leap about through rough ice like
a mountain goat, to say the proper word to the dogs should
they need any special advice, and to keep with the *komatik*
no matter where it went, up or down or between or under. It
was, as I have tried to say, a memorable expedition.

The moment the harness was put on Bobbie he cowered,
then bolted. Stopped with a violent jerk, he gave himself
over to a fit of howling. Somehow he got away from us and,
frightened by the long trace that rattled after him, he
bounded frantically to the outskirts of the post where the
trace became tangled among the boulders. He yelped and
whined dolefully. When I came up to free him he was quaking. I had to carry him back to the *komatik* in my arms.

The other dogs, strangely tolerant of the noise the youngster was making (eager, I fancied, that some interesting form
of punishment be administered), sat about in perfect order,
awaiting the command to go. Four times Bobbie was dragged
to his place in the team. Finally the whip-handle rapped on
the sled, the team leaped to their feet, and, at the joyous cry
of the leader, made off rapidly across the smoothly frozen
harbor. For a moment Bobbie thought he was free. With a
piercing *yip* he darted to one side. In an instant he was jerked
like a rag through the air, thumped soundly on the ice, and
dragged, a shapeless mass, for a quarter of a mile. The team
were eager to be off and resented delay. So dazed that he
scarcely whimpered, Bobbie was carried to his position once
more.

Again and again he shot off to one side only to be dragged
a hundred yards at a stretch. Becoming tangled among the
traces he was buffeted, pinched, nipped, flicked at with the
whip, and almost run over by the sled. He cut one foot on

the sharp ice. His tongue dripped blood. Once or twice he appeared to be at the point of collapse. Finally he began to see that so long as he kept in a certain place and ran with the other dogs, nothing very terrible happened.

All went comparatively well while we were on the smooth ice. Now, however, we were to cross part of the inlet where the ice had been broken and moved about by the tide. Bobbie and I had new facts to learn. We both came out of the experience a little the worse for wear. As for the minute details of Bobbie's difficulties (before he broke loose and ran home) I cannot say much, for I was too busy to watch him. As for me, I hurt my right wrist, cut three fingers of my left hand, wrenched my back, and froze my windward ear.

I take it you have never *komatikked* through rough ice. The great chunks are in no order at all, sticking this way and that, some of them on their edges, others upside down. You are travelling across a series of shelves, pinnacles, caverns, precipices and crevasses: a vast jig-saw puzzle heaped up ready for the ice-gods to play if they will. But the ice-gods never put their puzzle together; they only keep on scrambling the pieces, laughing when they see you and your dogs trying to get anywhere in the confusion.

The *komatik* heads every direction between straight up and straight down. The dogs don't mind much so long as the sledge doesn't pile down on top of them, breaking their backs or legs. They groan and whine, pulling as faithfully as they can under the circumstances. Sometimes you see the team, but most of the time you don't because they are under or behind some great chunk. The traces become tangled every few minutes, and they are forever catching on protruding points of ice. But you keep on, falling and stumbling, hanging desperately to the *komatik*-handles, and trying to be of assistance when you can.

We didn't get a seal. We didn't even shoot at one. Probably the *netchek* enjoyed watching us, their pursuers, from their cool divans out on the bay. We did see a snowy owl, big-eyed *Ookpikjuak*,[1] sitting on a hill-crest, apparently watching for lemmings. And we saw an *oogjook*-seal crashing its way up through ice two inches thick, trying to find a spot deeply enough frozen to hold him while he sunned himself.

I took it easy on the day after our "seal-hunt." The widow Kuklik brought me a pair of big sealskin mitts she had made. They were edged at the wrists with bear-skin.

Next day I was out again, walking on the new ice across to Seal Point. With some satisfaction I viewed what I could of the distant shore-line, remembering the long trips I had made, the mud-flats I had crossed, the streams I had forded, in getting to Seal Point in the summer-time. I began to understand why North Countrymen like the winter season.

I came upon the fresh tracks of a huge bear. Eagerly I followed them, scarcely believing that I might see the brute, but glad that I had my Krag instead of the shot-gun. I made the bad mistake of starting across a tidal pool among the rough ice, and fell through. I had a little difficulty in getting out, but did not feel disagreeably cold after the first frantic gasp or two, and decided to go on following the bear trail. But my sealskin *komik* went sloppy. Again and again I tied them on firmly as I could, but they sagged and sank down and flopped about my feet and would not stay put, and at last I gave up and started back to the post. I had learned an important lesson. Never again would I venture on crust new-formed on the tidal pools. Such crust might be dry and firm, but very thin. Light snow might cover it giving it exactly the same appearance as good ice. I would stay away from these treacherous tidal pools.

[1] White, Snowy, or Arctic Owl, *Nyctea nyctea.*

I came upon a seal-hole as I slushed and hobbled home, one of those round openings in the ice through which the *netchek* come up for a breath of air, or for a sunning, or for a nap. Such holes are about a foot and a half in diameter. Frequently they are rimmed with glass-ice that forms as the seals push water up ahead of them. There are so many of these holes that the animals may do their fishing almost anywhere and come up when they need air. Even in severely cold weather the holes are kept open because the seals come up so frequently. I suppose the seals find the holes by following the shafts of light that shine down through the dark water. A strange hunting-ground, this winter hunting-ground of the *netchek:* a world of dark water with a roof of ice three or four feet thick. I wish I could tell you how long a *netchek* stays under water between visits to the breathing-holes. But I cannot. I believe they can stay under a long time, for they not only catch, but they usually also eat, their fishy prey while under the ice.

I was peevish all afternoon. My frozen ear was sore. My back was sore. The nails of my big toes were miserably ingrown and out of shape as a result of these too-short *komik* I had been wearing; and here were these *komik* now, heaped on the floor, half-frozen, soggy, a shapeless mass. I put on dry clothes and tennis shoes and sat about grumbling. The worst trial of all was the triumphant return of Keetlapik, who killed a bear, that very afternoon, only about a mile beyond the tidal pool that had caused all my trouble. It was a monster bear. Keetlapik had killed it easily. We had bear steak for supper. But somehow that steak lacked a certain flavor it should have had for me.

. . . . .

Our drinking-water was melted from ice brought by the women from Lake Shoo Fly. I noted that there were some strange organisms in this water, tiny creatures that had the appearance of gray, semi-transparent, very young shrimps. We strained them out with cheesecloth. But they were very small and very sly and I have a notion that we consumed a great many of the more adventurous of them.

. . . . .

On the last day of October I saw several seals on the sunny ice, among them two great, black *oogjook*.

. . . . .

One of the dog-teams chased an Arctic fox across a lake. Had the nimble creature not changed its course we might have captured it. But it turned more quickly than the dogs could, using its long tail as brake and rudder, and bounded off to safety while we untangled the traces and set the *komatik* right-side up and made the dogs behave. Husky dogs are taught early in life that they must not so much as touch a fox in a trap. They can hardly be forced, in fact, to touch a fox-pelt or even the body of a fox that has been skinned. But they will chase a fox in the open, going wild with excitement.

. . . . .

I found two *tooktoo* [1] skulls, both with antlers. Since these antlers were very small we decided the caribou must have been females or young.

. . . . .

In the evening, far in the distance, I heard the howl of a wolf. It was a dismal sound.

[1] Barren Grounds Caribou, *Rangifer arcticus*.

CHAPTER XIV

Khianguliut

It was November, the season known among the Innuit as Khianguliut.[1] The *angenuk-nanook*[2] were fast asleep in their dens, dreaming such dreams as *angenuk-nanook* dream. The *angot-nanook*,[3] on the other hand, were on the salt-water ice, making their way to the far floe where they would spend the winter killing and gorging on seals. Nearly every bird and beast of the tundra had by this time put on its white winter dress. But *toolooghak*, the raven, was glossy black as ever, obstinate, headstrong *toolooghak*[4] that he was. And the small-footed lemming[5] still was brown, for he was to spend the white months under the snow.

On November 4, we had a cold snap. The thermometer dropped suddenly to thirty below. The wires of the radio aërial hummed and whined all day. The water that filled the hollows in the rough shore-ice at high tide steamed and smoked, then "cooled off" as a scum of crystal formed.

One of the youngsters, who was drinking from a granite-ware mug that had lost most of the granite in the course of its eventful Arctic existence, suddenly found himself frozen to the rim. He tried to pull himself free and jerked part of his cheek off before the women came to the rescue.

I asked Sam whether the Eskimos ever caught weasels on

---

[1] The time when ice forms all round the ocean-shore.
[2] Female polar bears.
[3] Male polar bears.
[4] Northern Raven, *Corvus corax principalis*.
[5] Back's Lemming, *Lemmus trimucronatus*.

grease- or blood-covered knife-blades set out in cold weather. I had heard that weasels would lick the grease or blood and in so doing freeze their thin tongues fast to the metal. But Sam said that the Eskimos now living on Shugliak paid no attention to weasel-catching for the pelts were worthless "in trade."

John Ell told us he was going to Native Point to get the widow of One-eyed Joe. One-eyed Joe now had been dead a year. John wanted to care for the widow until she got a husband, for John had more food and clothing than he and his immediate family needed, here at the post.

.   .   .   .   .

Sheeloo and Tapitai came in from the Koodlootook River country, announcing that they had shot four caribou. Scarcely had they finished their trading when Shookalook, that tall, curly-haired Portuguese of an Aivilik, came in from Duke of York Bay telling us loudly of the thirteen caribou he had killed and bringing us the thirteen frozen tongues. I liked Shookalook exceedingly well. Whenever he came to the post he asked me to play the organ at the Mission. Something about Shookalook's lambent eyes made me think he might be able to sing *"Si alguna vez en tu pecho, Ai Yai Yai!"* or *"La Golondrina"* with genuine romantic fervor. There was the air of *un caballero* about him. He would have been transported by a bull-fight.[1]

.   .   .   .   .

I was amused at the behavior of a flock of ptarmigan that had occasion to walk through some thinly encrusted snow. They forged straight ahead, taking dainty steps as ptarmigan do, until the crust broke in, whereupon they fell flat on their

[1] Shookalook, poor soul, now is dead. He was drowned in the year 1931.

faces, stumbled about as if embarrassed at such an experience, and finally righted themselves, found firm footing, and shook the snow out of their eyes. When I laughed aloud at the birds, their faces took on the most absurdly quizzical expressions.

. . . . .

The pup we called Ginger Junior (a dirty-orange young scamp) was not walking correctly. Something was wrong with his spine or pelvis. Some woman probably had struck him too hard with a club.

. . . . .

The lazy Kungualook, who never caught a fox nor shot a seal nor even carved out good *tooghak* [1] dog-harness buttons, came in one evening while Sam and Jack and I were playing Hearts. He did not knock and this lapse Sam did not like. The Eskimo said not a word, but a noisy odor accompanied him. He probably hoped we would give him some food. But he did no begging and grinned sheepishly when we looked at him. Not being offered a chair he squatted on his hunkers, watching us furtively, and finally picked up a *Literary Digest*. He left after a time, mumbling that he thought it now time to go somewhere else. Sam said he was a ne'er-do-well, that he hunted a little but never shot anything, and was always out of clothes and ammunition and food. He was of the Okomiut. He was the post's garbage-supervisor. In summer it was his task to pick up and carry away the larger stones that had gathered during winter in the formalized gravel-plots surrounding the house and store.

. . . . .

[1] Walrus-tusk ivory.

The daylight hours were now so few that there was little time for hunting. I unpacked the bird-skins I had collected and salted-down at Cape Low and Seahorse Point and began the laborious process of fat-scraping, rinsing and stuffing. On the brighter days I worked at a series of water-color paintings.

· · · · ·

Since nearly all the fox-pelts coming in these days "for trade" were prime, Sam and Jack and I decided we would put out trap-lines. We planned first to scatter bait in likely places between the post and Poorhouse Hill. Traps would be set once we learned where the foxes were. Jack and I got *komatik* and dogs and rotten fish ready. But while we were inside the house finding our rifles the dogs dashed off after a team that had headed long since for Keetlapik's encampment at Koodlootook River, seven miles to the west. They were not caught until noon. Young Noah (I don't remember the lad's Eskimo name) brought them back. Punishment was severe. Whips lashed fiercely. There was more than the usual howling. You'd think the dogs would know better than to dash off that way without a driver; but sometimes they don't. The *komatik* was considerably damaged.

· · · · ·

I took the skinless body of an owl out to Kutchoomaitook, the nervous bitch that had deserted her pups. She rushed up, grabbed the carcass and made off to the shelter of a rock. Here she dropped her prize as several dogs ran up. A fight started, each of the brutes believing one of the others to be making off with the owl. Meanwhile the bitch stole back and chewed away to her heart's content.

· · · · ·

It was becoming steadily colder. I happened to notice, one day, as I was returning from a hunt, that what I could see of my nose was waxen white. I rubbed it with my warm hands.[1] It did not go white again that day. But next day the skin cracked and scabs formed. This, I should say, was the beginning of the "long, dreary winter" for me.

[1] Never rub a frozen nose with snow in the North Country. It isn't done.

CHAPTER XV

A Dissertation upon Huskies

A TWO-DAY blizzard shut us in. My nose was unpleasantly sore. I decided to take a vacation and do some writing. I wrote a sort of paper upon Husky dogs that I entitled *King-mik*. I wrote this paper upstairs, straight above the big kitchen stove. For light I had that tall kerosene lamp with its frail net burner that was sure to crumble to bits whenever I had an especially good idea. Here is the dissertation, almost exactly as I wrote it.

Kingmik

Kingmik is the Eskimo noun for a Husky dog; a *king-miatsuk* is a puppy; an *angenuk* is a bitch (*angenuk* is also the word for an Eskimo woman); an *ittuk* is an old dog, and so on; but whether he be *kingmik*, *kingmiatsuk*, or *ittuk*, a Husky's external, physical life is little more than a succession of bloody battles relieved by an occasional good howl, a gorging of blubber or rotten meat, a sleep in the snow, and hours of *komatik* pulling. Of his mental and spiritual life I have learned little, though I have noted that he has an active brain, and I am in no position to deny that his individuality may reflect a soul quite as perfectly developed as my own.

Through a clear corner of the frosted storm-window I can see the post-dogs now. They are well fed and there is, for the moment, no occasion for fighting. Tweed, a massive leader-dog, sits in the snow contemplating the landscape, his small eyes almost hidden in the grizzled hair of his face. He sees a mound of snow, a tide-raised pinnacle, a distant patch of exposed rock: some object worthy of a moment's careful study. He lifts his head, sniffing the air with his black-tipped nose. John Ell is passing by.

"What does the dog smell, John?" I ask.

John answers promptly, perhaps with essential accuracy, "She not smell. She thinking of animal!"

Tweed's accomplice in crime is a big brute whom I call Play Boy. His Eskimo name I cannot write down; it means "He Likes to Break Up Something."

Then there is Koopaitook, the white-chested one with the bared fangs: snarling, cowardly Koopaitook, whose name means "His Hair Is Not Parted." And Blackie, a low-hung animal, who is Koopaitook's constant tormentor. And Ginger, the dirty-orange, flop-eared arsenic-eater. And poor old Sailor, the vanquished leader. And Kutchoomaitook, slender as a collie, with lean and wistful face. Her name means "She Is Willing to Go Ahead." And young Bobbie and his sister, Annie, who are really too young to count, save when you step on them just outside the door, or when they steal something.

There is a difference of opinion among men of the North Country as to the Husky dog's ancestry. Some believe him related to the timber or the Arctic wolf. Some aver that the wary wolves of Shugliak are Huskies gone wild. Many a Husky might, indeed, pass as a wolf. But the black animals, and those with plumelike tails that curl over their backs, remind you of such breeds as the Chow, the Newfoundland, the Labrador, the Spitz, and the Samoyed.

Blind but well-haired, snub-nosed and short-legged, *Kingmik* comes into the world at any time of the year, even in the dead of winter. He grows rapidly if the Eskimos, the older dogs, and the relentless elements grant him existence. On the eighth day he opens his eyes.

If he survives the vicissitudes of babyhood he fares forth with his mother to learn what every pup must know. The rules of his world, he soon finds, are about as follows: First (and this is most important), if you come upon anything edible, be it sealskin boot, walrus-hide whip, grease-soaked rag, blood-covered chip, mouse, fish, seal or whale, eat all of it *if possible*, and as soon as possible, for you may never have the chance again. Second, if a dog bites you, you need not necessarily bite him in return, for you may waste time in chasing him. Bite any dog you can, and bite him hard, then duck for your life. Third, howl upon all possible occasions. Fourth, in choosing a place to nap, select the most frequently used thoroughfare you can find. This is sure to lead to admirable trouble.

Kingmik soon learns his place in the social scheme. If rotten *netchek* is being fed he must hang about the caudal periphery of the feast waiting opportunity to sneak in between somebody's legs. He will be bitten, of course, and he will probably get little more for his pains than an agonizing sniff; but he may snatch a stray strand of intestine, a chunk of fat, or a blood-covered stone. This prize he carries to a safe distance, where, in as meek a manner as possible, he may eat. If he steals a sizable chunk of flesh he pays for his insolence. An older dog overtakes him, bites him hard three times, on the head, in the ribs, and the other spot is optional, and eats the prize on the spot. *Kingmik* now moans hideously, as if he has been eviscerated or had his back broken. After the principal spasm of moaning is past it is permissible to moan considerably longer; then it is best to forget such minor matters as eviscerations and return to the feast.

If a whole *netchek* is being consumed, *Kingmik* may actually crawl inside the ribs of the vile carcass and there gulp away at random, immune for the time since the other dogs are busy with the outer muscles and flippers. But woe to him if he attempts this daring feat and fails!

I am not sure that *Kingmik* purposely gets as dirty as he can at a seal-feast. But he acts as if there were an added joy in wallowing in fat and blood, and when, round-bellied after the gorging, he lies down for a nap, his matted hair is redolent of all that is unlovely in the North Country.

During winter, especially when the team is working, the dogs are fed every other day. The winter-born puppy therefore fares well. In summer the dogs are fed virtually nothing and the puppy, his rapidly growing body needing nourishment, sometimes finds it difficult to keep alive. He wanders along the beach searching for dead fish cast up by the waves. He haunts the doors of the *tupek*, eager to lick, gnaw at or swallow almost anything that is thrown out. Perhaps he grinds away at a discarded boot, or an old mitt.

One of the young dogs at Cape Wolstenholme, so Sam tells me, once became so violently ill that everyone expected him to die. After a spasm he regurgitated a pocket-knife, blade open. The blood-covered knife had been irresistible.

Kingmik is destined either to take a place in the *komatik*-team or to be killed and skinned. The Eskimo so depends upon his dog-team that he values a promising pup highly. *Kingmik* is broken to the harness when he is about half-grown.

The harness is fashioned from narrow strips of *netchek*-skin. Slipping snugly over the head with ample openings for the front feet, it is attached by a snap at the back to a long trace of *oogjook*-skin that leads to the *komatik*. Here all the traces, which terminate in walrus-ivory "buttons," are threaded upon a tough thong that is bound securely to the front of the sled. The traces are of different length so as to permit the several dogs to run as comfortably as possible, in fan arrangement, ahead of the *komatik*.[1] The leader's trace is, of course, the longest.

The sled itself usually is ten or twelve feet long and three feet wide with high, strong runners. Frequently there are handles at the back that the driver may grasp when he is steering from the rear, and that serve as support for the load. *Komatik* were formerly made of whalebone; today wood from the Outside is used. The sled is often heavily loaded. Each dog is supposed to be able to pull about five hundred pounds. Teams at Shugliak usually number from seven to fourteen dogs.

The leader of the team is a privileged character. He bounds back and forth since there are no dogs beside him. He drops back among his comrades to shout encouragement, to settle a dispute, or to touch noses. If he misbehaves and is whipped for his waywardness he moans as if his agony were of soul rather than of body.

Kingmik is a born hunter. His keen nose may detect a band of *tooktoo* miles away. His master may search for days in finding a good *netchek*-hole, but *Kingmik* finds three or four holes without difficulty in crossing a stretch of frozen salt water. The Eskimo permits his dogs to wander where they will in hunting seals, and the dogs, realizing that a capture

---

[1] This arrangement is the only arrangement ever effected on Shugliak. In certain other parts of the North Country the teams are arranged tandem.

means a meal for them, are as much interested as their master in the success of the enterprise. On the alert for a scent the moment they reach salt-water ice, they zigzag along, sniffing at the cracks and racing in high glee when they catch a strong lead.

Kingmik helps his master catch foxes. The trap-line must be visited regularly, for wolves will eat the foxes that have been caught and almost never step into a trap themselves; ravens will steal the bait; wind will bury traps deeply with snow; and furthermore, when a trap has caught a fox, that trap will catch no more until it is reset. Traps may be set miles apart, and the Eskimo who puts out a line of a hundred or more traps must have *Kingmik's* help.

The battle with *Nanook* is the peak of *Kingmik's* career. Here he meets face to face his most glorious foe. The size of the bear, the savage growling, the lightning-quick blows of the mammoth forepaws incite the dogs to a frenzy that is sometimes overzealous. Many a Husky has come an inch too near, waited a fraction of a second too long, or had the misfortune to collide with an eager comrade, and has paid the penalty with his life. For the blow of *Nanook's* forepaw is mortal. The dogs may slash and gouge as they will at the shaggy hind-quarters, doing little harm, for the hair is long and thick; but once they jump for the bear's face, throat or ears, they are courting death.

Naïvely the Aivilik Eskimos tell the story of a man who hunted *Nanook* on Shugliak a long time ago. Being poor this man had but one dog. He was very fond of his *Kingmik*. In battle with a huge bear, *Kingmik* was killed. The man crouched on the ice and mourned. *Nanook*, beholding from afar the grief of the stricken hunter, walked over to the dog's carcass and lay down beside it. The bear's spirit entered the dog's body. *Kingmik* rose with a joyful whine and returned to nuzzle his master's hand. Together they went back to the bear and skinned it.[1]

So inseparable are the Eskimo and his dogs that many legends concerning the origin of the Innuit include *Kingmik* at the very beginning. Some stories relate even of the mating of Eskimo and dog and the consequent peopling of islands never before inhabited.

But whatever you may remember about *Kingmik* and his ways, be sure to remember his howling. For the Husky dog chorus is the national anthem of the North Country.

[1] This story reflects the characteristic high regard the Aivilik Eskimos have for the polar bear.

CHAPTER XVI

The National Anthem of the North Country

MY CALLING the dog chorus the North Country's national anthem is not altogether an attempt to be clever. It *is* a little funny, the way the dogs rise to their haunches all over the post whenever the howling begins, some of them getting up briskly, for all the world as if the opening strains had roused deep patriotism, others squirming and writhing and finally rising as if they were determined to see this thing through no matter how boresome it all might be. It is funny, too, when you observe that the dogs aren't really singing together. One dog sings this, another dog sings that, each of them inventing his own tune as he goes along. The general effect, you cannot help thinking, is not so different from that of an average American audience singing "The Star-Spangled Banner." Every man for himself is the big idea; personal liberty at any cost, and the devil may care for the rhythms and harmonies!

At the post you hear the dog chorus at least three times a day: at seven o'clock in the morning, at noon, and at six o'clock in the evening when, for a period of about thirty seconds, the bell at the Roman Catholic Mission rings. At the first clanging note, sometimes indeed before the sound has reached human ears, every dog in the place is on the alert; and scarcely has the second note pealed forth when from some quarter the chorus begins. The physiologist may tell you that the dogs howl because the ringing of the bell annoys them, disturbs the mechanism of their ears. The psychologist may tell you they howl in response to a certain challenge the bell hurls forth. My explanation is simpler: the dogs howl because they want to, because it's fun, because it's something to do.

Every dog-fight winds up with a chorus, of course. Now the performance is a sort of formal expression of sympathy for the comrades who have lost teeth or eyes or pieces of ears. It is as if the dogs were singing to one another: "We have fought together nobly. It was a glorious fray. There was much bloodshed and pain. We enjoyed it very much. Fighting is part of our being. But we are weary with it now. We are all weary with it now. How solemn a matter life is, after all!"

One or two whining voices start. There is a querulous yap or two, a low growl, the complaint of a puppy, as the animals lift their heads. Now a deep, velvet-smooth voice takes the lead, swelling ever so gradually both in tone and in volume, to a mellow wail that persists through the cacophony soon to develop. High, sharp notes are sounded. The puppies, bursting into falsetto moans, furnish a primitive rhythm. There are sopranos clear as the Mission bell itself; contraltos, rich, well-rounded and even; cavernous basses; ear-splitting contra-tenors; voices that sound like wind instruments of an unearthly sort. The strange chords are not, I suppose, purposely harmonized; some of them are unsteady, weak, and shrill; some grate upon the ear; yet there is harmony—a harmony that somehow charms.

The essential spirit and intent of the dog chorus is old as the race of wolves; its musical pattern, however, is decidedly modern. Its lack of obvious melody, its avoidance of all that is trite, its skilful weaving together of unrelated keys, its refreshing discords—all this would delight a Ravel, a Prokofiev, or a Debussy, even though its morbidezza intricacies might baffle him.

You cannot analyze this music too technically, for it is not that sort of music. But you come in time to sense that it mysteriously educes and fuses the spiritual values of the North Country as a national anthem should. It comes to symbolize all that is ineffable; all that is defiant and untamable; all that is lovely and remote; all that is transformed into crystal, frozen into silence, yet throbbing with brave and patient life in this heaven-bound Arctic world. You nearly always smile, or laugh out loud, when the anthem begins, for the sound indubitably is funny. But it is also *not funny*. Your laughing may cease as you suddenly find yourself embarrassed, confused, lonely. It is all very funny until you realize that it is *your* anthem as well as their own.

If the dog anthem is thrilling at the post at the white-gold hour of noon, it is thrice thrilling at night in some little, remote encampment out at the floe-edge. Especially at night when it is inspired not by any bell, not by a fight, but by the spectral aurora borealis. The writhing lights rouse, perhaps even frighten the animals, and the chant ascends. Now there is a quality of restraint in the deeper voices, an acknowledgment of respect, of awe, or of fear, and a hint of supplication in the shriller tones, as to the God of the Cold Night. There is no falsetto moaning of puppies; for the nocturnal chorus at the *sheenah* [1] is a ritual of mature souls.

When your head turns at the first whine and the nape of your neck tingles at the rising wail, you are experiencing an emotion that may be basically fear. Just as the eerie melody is, itself, a response to impulses deeply rooted in the wolf ancestry of the dogs (I see I am turning psychologist now!), so your enjoyment, which borders on hypnosis, probably harks

[1] Floe; ice-edge.

back to the dread your distant forbears had for the beasts that carried off their children by night. The age-old chorus dissolves the lowly *igloo* and their dim *koodilik*-lamps into thin, frost-hung atmosphere; the *komatik* and the civilization they represent fade into a misty void; the world turns back an eon or two, and all of you, dog and man alike, acknowledge for the moment the limitations of your bodies and souls, your dependence upon the wishes of the Great Powers that control you both. The dog-wolf fears the erratic splendor of the aurora; or, stimulated by its brilliance, sounds the ancient hunting cry as a challenge to Starvation and Death. The many voices join in expressing a conviction that the pack is more powerful than the solitary animal. There is the underlying assurance that the wind-borne tones will make many a tiny beast out on the tundra quiver in mortal terror; but there is, just as definitely, an acknowledgment that the days of the lordly wolf are numbered.

And you? Warm in your caribou-skin sleeping bag, modern representative that you are of another ancient race of beings, you pass a cozy hour in thinking; using your human brain that is, you have been told, so much more highly developed than the brain in the "lower" animals. You recall that you are a sort of god, but "little lower than the angels"; that your kindred have all but conquered the universe; that you long since have fettered the beasts of the field, subjugated the ocean, harnessed wind and waterfall, and reared your temples of industry into the face of the sky. "There is nothing I cannot do, or learn to do, if I but will to do it," you say to yourself. "And there is a Heaven awaiting me at The End."

The End! How glibly the idea is thought! You stop thinking, suddenly smitten with dread. So accustomed you are to acknowledging your utter impotence in dealing with or even comprehending Death that, in the high hour of elation, you prefer to shut the thought of It from your mind. Most tractable, most comfortingly adjustable, is the human intellect when moments fraught with such calamitous possibilities present themselves.

Through a crack in the snow-house wall the nebulous, shifting Aurora seeks you out. Half an hour ago she was only the northern lights. Now she is a goddess celebrating festival. With a faint sound that you cannot help thinking a derisive hiss she flings her luminous mantle across the sky, draws it back upon herself, flings it forth again, taunting you. "And what do you know about me, Great Man, you of the Superior Race? You have refused to think of Death, for you find it uncomfortable to face the Unseen, the Unknown. You prefer to think of yourself as the Baffler, not the Baffled. Here am I, probably not so very far above you. You see me plainly almost every night. What do you know about *me*? Surely I am not Death, am I? Or can it be there are other matters than Death you prefer not to think about? I might, you know, be very useful to you!"

Hurled utterly from the peak of self-glorification, you find solace in the companionable howling of the Huskies. Once again they are singing your anthem. The mellifluous, long-drawn-out discords die away in the blue darkness as the friendly animals lower their heads, turn about once or twice in observance of another olden custom, and curl up alongside the snow-wall only a foot or so from you.

"Well," you say in resignation, as you begin to yield to the physical need for sleep, "the dogs are a friendly lot. They doubtless know a very great deal. But they can't know much more about the aurora than I do!"

CHAPTER XVII

Teregeneuk, the White Fox

You have heard a good deal about foxes. You remember the fox in the fables of old Æsop, and the sly, clever Reynard of mediæval tales. You recall that there are red foxes and gray foxes in the United States, and that the more southerly of these, the gray fox, sometimes climbs trees. You know there is a fur called Arctic Fox, too, for you have seen it at the Opera. But knowing, as you do, that manufacturers of furs are given to offering all manner of rabbit and alley-cat skins under such fetching names as Himalayan Lapin and Senegal Civet, you wonder if Arctic Fox isn't albino muskrat or raccoon, or just any old fox that has been run through some heroic process of bleaching.

The Arctic fox is a real fox. The Aivilik Eskimos call him *Teregeneuk*.[1] He comes in two color-phases in winter: the white; and the so-called "blue," which is not really blue at all, but a handsome shade of smoky gray. In summer he is dull brown above and yellowish underneath. In some parts

---

[1] White or Arctic Fox, *Alopex lagopus*.

of the Far North "blue" foxes are common, commoner, in fact, than the "whites." On Shugliak, according to records kept at the post, one "blue" is caught to about every one hundred "whites."

The Eskimos think of *Teregeneuk* as a sort of downy hub about which the wheel of the White Man's Civilization moves. There is a trading-post and Sam Ford and Jack Ford and the great steel-plated *Nascopie* that comes every year; there are rifles and flour and tea and gramophones and motor-boats and alarm-clocks and trinkets and candy and silk handkerchiefs and blue suspenders for trade, all because of *Teregeneuk* in his winter coat. Whales have become rare in Hudson Bay, so the White Man says little about whales nowadays. Nor does he make special request for carvings of *tooghak*-ivory, or for bags made from the skin of loons' heads, or for bone-handled or antler-handled knives, or for *tooktoo*-hide dickey-coats and trousers. But he does ask and continue to ask for white fox skins. "Get me all the foxes you can get! Now that you have got these, go back and get more! Here are traps. Hurry now! Be sure to get foxes that are all *kadlowktok*.[1] The spotted ones, the ones with patches of summer fur, will not do for 'trade.'" So speaks the White Man. And the Eskimo probably wonders, in his manner of vague wondering, if the White Man of the South Country makes all his finer clothing out of fox skins, or if he makes sleeping-bags out of them, or if he decorates his *igloojuak*[2] with them, or if he uses them in strange religious rites.

Teregeneuk is, you perceive, Shugliak's most important fur-bearer. *Teggeuk*, the weasel, doesn't count, in spite of the high-soundingness of the word *ermine*. Polar bear-skins are not in great demand. Nobody cares a whoop about *Ookalik*,

---

[1] White.
[2] Literally "big *igloo*." The word has come to signify a white man's dwelling-house.

the hare, though his fur is marvellously deep and soft and warm in winter. And little platinum-gray *Uvinghuk,* the big-footed lemming,[1] is so difficult to capture, and his lovely pelt is so fragile and hard to match that nobody dares yet to try to make him fashionable. So everybody traps foxes. The Eskimos of Shugliak sometimes catch fifteen hundred foxes in a year.

Obviously, since you are on the island for a whole winter, you must learn to catch foxes. Since it is what is called a "good fox year," you may set your traps almost anywhere, even near the post. It is a good idea to watch for "signs," of course, and to put the traps in places where you know the foxes recently have been.

You will note fox-signs especially along drifts in which lemmings live: faint scratchings and clean-cut claw-marks, pale-yellow stainings, and deep shafts that reach down to the very heart of the drift. Near these shafts, if the wind has not been strong, you will find the scattered remains of lemming nests and tiny spots of frozen blood.

You have your traps in a sealskin bag. Through the ring at the end of each trap-chain you have fastened a stick of wood about two feet long. This stick is to help in anchoring the trap.

You trample the snow down firmly in the spot you have chosen. Here you lay the trap. Having fetched a good-sized boulder, you drop this heavily on the stick of wood. The trap is now anchored, for the snow will freeze firmly about the wood and the end of the chain in a surprisingly short time.

With your snow-knife you quarry from the drift a block about two feet square and eight or ten inches thick. This block you set, broad side down, against the boulder, with the trap-chain between boulder and snow-block. Having set the

---

[1] Richardson's Lemming, *Dichrostonyx rubricatus.*

trap proper, you lay it, open-jawed, on the block. With the point of your knife you make an outline of the trap. Following this outline you scoop out a basin that will just hold the trap. In this basin you sprinkle bait: bits of rotten seal-meat, crumbs of cheese, or some "Florida Water." [1]

Having placed the trap in the basin, you cover trap, bait, basin and all with a thin, firm slice of snow that you scrape carefully down until you can almost see the trap through it. Perhaps you scatter more bait here and there by way of making the place alluring. Then you walk away, making no more footprints than are necessary. You need not be especially careful, for *Teregeneuk* is, as a rule, more curious than suspicious.

Sam showed me how to set a trap. I chose the country east of the post as my trapping-ground and set to work. Within a week I had put out a line of about thirty traps. But a long time passed before I caught a fox. Day after day I came in with tales of traps that had been sprung, traps that were buried with snow, traps that held frozen weasel carcasses and hares' toes. I began to think I was not much good as a fox-trapper. My only consolation was the realization that neither Sam nor Jack had caught more than one or two animals.

I learned a good deal about *Teregeneuk* and his ways. With my binocular I saw him trotting about listening for squeaks and sounds of burrowing beneath him. I saw him start to dig, then stop to listen. I saw him shove his nose, terrier-wise, into the snow, there to sniff and snuffle. I saw him bound to another part of the drift and start another burrow. Finally I saw him set to work in earnest, sending a steady spray of snow out back of him: at last he had decided upon the location of the lemming nest and was on his way down to a meal.

[1] Perfume.

One day I surprised *Teregeneuk* at his lemming-catching, and pulled him out of his burrow by the hind feet. He was very fierce and very quick. Before I knew what was happening, he had bit through my mitt, severing a small vein. He was an exquisitely beautiful creature, with golden-brown eyes and only a wisp of black hairs at the tip of his thick tail.

I trapped a good many foxes once the tundra-gods decided to smile upon me. Most of them were caught by two or three toes or by a front foot. Since I was interested only in specimens, and not in pelts for trade, I liberated many animals that I felt I did not need. These ran off a trifle stiffly; but they ran as if they thought running far preferable to dying. And I think their injuries did not greatly inconvenience them, for the Eskimos were constantly bringing in pelts that were minus toes and feet. These pelts were from animals that obviously had been in perfect health at the time they were caught, their trap-wounds completely and neatly healed.

The Eskimos have all the steel-traps they can use, nowadays; but they sometimes repair and use the stone traps that the Shugliamiut built in olden times: hollow piles of boulders four or five feet high that are to be seen here and there all over the island. Snow buries these stone traps to be sure, but the hollow spaces inside remain snowless, and *Teregeneuk*, having made his way in through one of the small openings at the top, finds he cannot leap out.

The Aiviliks sometimes catch foxes also in wire snares, or in a barrel with the lid revolving on an iron rod. *Teregeneuk* jumps on the barrel; one side or the other of the lid suddenly sinks; and *Teregeneuk* finds himself standing beside a fine, rotten fish, inside the barrel. He does not sense that anything is wrong until, having eaten the fish, he tries to get out.

The most serious rival of the fox-trapper on Shugliak is the Barren Grounds wolf. On the mainland to the west, the wolverene is common enough to be nuisance; but the wolverene does not inhabit Shugliak.

How can the Shugliak Eskimos continue to capture so many foxes each winter without exterminating them? This is an interesting question. An adequate answer must take into consideration the desultory trapping methods of the Innuit; the tendency of the fox to range widely as a species; the abundance of lemmings and other fox-food during an average winter; and the absence of what are called "natural" enemies, on this nineteen-thousand-square-mile island. But more important than these several factors is yet another factor: the size of *Teregeneuk's* brood. A mother fox will sometimes give birth to as many as twelve or fifteen young. If the food supply is adequate for all these cubs most of them will reach maturity, for there are not many fox-cub destroyers on Shugliak. The presence of all these young animals in their first winter coat swells the catch tremendously. Furthermore, were not some of this vast population of foxes to be caught in traps, *Teregeneuk* might conceivably become so numerous that he would destroy his own food-supply and find himself faced with starvation.

I have not seen many fox-cubs on the tundra. But I have heard them chippering and growling in their dens, and I have counted the heaps of ptarmigan wings that litter their playgrounds. *Teregeneuk* kills a great many *ahigivik* during spring and summer, if these grouse are common.

Teregeneuk is scarcely comparable to the famous Reynard in cleverness and wit. I should, in fact, call *Teregeneuk* a somewhat stupid and gullible beast. But wait until you hear some of the Aivilik tales about him and you will decide that foxes are somehow destined the wide world over to embody the traits of the foxy.

CHAPTER XVIII

Tragedy

IT WAS the evening of November 15. We had had a busy day. Sam and I had walked to Seal Point to set fox-traps, talking, as we went, of the trip I was to make with the Eskimos to East Bay. Jack had gone to the head of the inlet, visiting his traps and bringing back one fox. He had taken the young bitch, Annie, on her first *komatik* haul. She had been exceedingly difficult to manage, had finally broken loose from the sledge, and had got herself caught in two fox-traps on the way home.

We were seated in the "parlor" talking about our experiences when we heard footsteps and a quiet knock.

It was Father Thibert from the Mission. As he entered the room we could see that his face was serious. Wasting no words he told us that he had just journeyed in from Cape Low; that he had visited the Okomiut encampment beyond Munnimunnek Point; that he had sad news for us. Two of the hunters at Munnimunnek had drifted off on the ice. And they had been given up for lost.

It was not a long story, only a few words, most of them verbs and nouns. Stories of death need no adornment.

"Do you remember," Father Thibert began, "that we had a gale from the northeast about ten days ago?"

Yes, we all remembered.

"One of the Okomiut hunters was out that day, following a bear through the ice. He was about a half-mile from the encampment beyond Munnimunnek when one of the women at the camp suddenly saw that a great mass of ice had broken free and was drifting rapidly out into the bay. Already there was a wide channel separating the drifting mass from the *sheenah*, the solid ice-shore.

" 'The ice has broken off!' she screamed. 'The hunter of *nanook* is drifting out to sea!'

"All the Eskimos came out of their *tupek*, stricken with terror. One of the men rushed across the ice to his canoe. He was only half-dressed; he had no outer *kooletah*, no *komik* over his caribou-skin stockings. He slid his canoe into the water, snatched up two paddles, and shot off through the choppy waves. Every moment the channel was widening and the waves becoming rougher.

"The Eskimos ran out to the *sheenah* to watch. The ice on which they were standing might have broken off too, but they didn't think of that. They were thinking only of the *nanook*-hunter and of the man who was trying to rescue him.

"They saw the hunter struggling back across the ice. He couldn't run because the wind was blowing so hard and because the ice was so rough.

"The canoe got across the channel quickly, for the wind blew it along as if it had had a sail. But managing the canoe on the windward side of the drifting ice was difficult because the waves were so rough.

"By the time the canoe got out to the ice, the *nanook*-hunter was only about halfway back. And all the time the channel was growing wider and wider and the water rougher and rougher.

"The people on the shore couldn't see very clearly by now. They put their hands up to shelter their eyes.

"Finally the *nanook*-hunter got back to the canoe and the

two men managed to get in. They struggled along, trying to make headway. They succeeded in paddling out a little distance, then had to turn back. They tried again but had to go back. They tried once more and then appeared to give up trying. The people couldn't be sure what had happened. Some of them thought they had seen the canoe turn over. Others were certain they could see the two men standing together on the ice. But they didn't see the canoe again.

"The people would have sent another canoe out, but they didn't have another one.

"By this time the ice was so far away it was just a thin line of white along the horizon."

Our little parlor was a silent room. We looked at one another realizing full well there was nothing we could do. The two men had drifted away ten days ago. They had not been seen nor heard from since. It was all so like a story to me, so unreal, as if I had been only reading. Yet there was Father Thibert, leaving us now, his face so red and weatherbeaten, his eyes so very sad. And there was Sam, looking at the floor. And Jack, with not a hint of the smile that usually played round his eyes and mouth.

We talked a little, thinking as we talked of that loyal man who had paddled across the ever-widening channel in his canoe. No sealskin boots. No outer *kooletah*. We thought of the *nanook*-hunter. One rifle. A few cartridges, perhaps.

Where were the two men now?

Could they possibly be alive?

I could scarcely sleep that night. Out in the darkness that was my room were the figures of two men. The figures were standing now; they were hunched over now; now they were lying on the drifting ice. Out in the blackness and coldness that was Captain Henry Hudson's bay. I wondered if they might drift across to the far western shore. Or if a southerly wind might blow them back again to Shugliak.

We never saw them nor heard of them again.

CHAPTER XIX

Little Peter

IT IS strange that I have got so far in my tale without telling you more of Little Peter.

I had my first glimpse of this six-year-old, full-blooded Aivilik lad about five minutes after I set foot upon Shugliak. He was standing, as if transfixed, near that filthy half-grown polar bear, down along the shore. When Peter saw me that day he did not smile. Like his numerous child companions he regarded me dubiously. When I took a step toward him he scampered off among the stones agile as a hare. From a safe distance he eyed me, his black hair dishevelled by the wind, a finger stuck into one corner of his mouth. I smiled but did not draw closer. It occurred to me that Peter might embarrass all of us by bursting into tears.

By November Pete and I had become good friends. He even had told me in his quaint, not to be pronounced and scarcely to be translated baby-Eskimo prattle, that he no longer felt "shy" when in front of me.

Pete's given name, his personal Aivilik name, that is, was unmentionably obscene. Why his parents ever gave him such a name is beyond me. About such matters they were, I decided, either delightfully, refreshingly frank, or hopelessly sordid.[1]

Pete was seven hundred and eighty millimetres tall. You

---

[1] I extend to myself no compliment when I make this statement. My questioning their frankness probably furnishes proof of my own low-mindedness.

will think this an odd way to record an Eskimo boy's height; but Pete preferred to be measured as he had seen lemming and weasel and fox specimens measured, with that zippy steel metric tape that pulled shut when the spring was touched. So it would be breaking faith to say anything about feet and inches here.

Pete stood very straight. His legs and arms, while short, were strong. His flat face, stubby nose, exceedingly bright brown-black eyes, and dark skin all made me think of a little Jap, though his eyes were not slanted.

Of what the Outside World calls cleanliness, Pete knew next to nothing. He never, so far as I know, washed himself, nor combed his hair, nor brushed his teeth. The servants' house where he lived we must call *dirty* for crying want of a more accurate word. Tools, chunks of wood, steel-traps, dog-harness, harpoons, boots, mitts, and seal-meat lay in heaps here and there. Frozen foxes, sometimes dozens of them, hung from the ceiling, awaiting the hour when, properly thawed, they would be skinned. Fox skins lay everywhere, in boxes on the floor and in heaps in the corners: some white and downy as powder-puffs, others bloody and greasy and befouled and ill-smelling. On one of the three small cots in the principal room stood a hand sewing-machine. Among the cans, pots and comestibles on the table was a small gramo-phone. Binoculars in leather cases, rifles and cameras hung on the dingy walls. Near the doorway a huge caribou-skin was stretched on a rectangular frame. Since the double windows were not only dirty but also deeply encrusted with ice, the light was dim. Everything but the recently trapped foxes was filmed with dust or smoke. There was a heavy odor of seal-oil. Women were squatted on the beds sewing clothes, mending boots, or scraping fox skins, using their teeth as well as their fingers. Old Angoti Marik sat on a box, endlessly skinning out fox-feet, using a small, sharp pocket-knife.

Pete's waking thought was always of food. *"Kahpoonga!"* [1] he would shout; and everyone would know that he was well. His "wild" relatives out at the floe might breakfast on *netchek*-blubber and *tooktoo-quak* and tea; but Pete could, if he wished, have jam or slap-jacks or condensed milk for his meal. He lived, you see, among Shugliak's elect, where the servants took the left-overs from the Chief Trader's table and where rations were regularly dealt from the store.

Breakfast over, Pete was shaken and jammed into his snug winter rigging, for he preferred to play outdoors save in the worst gales. Over his woolen underclothing he wore thick shirt and trousers. On his feet were two or three layers of boots, the innermost of caribou-skin with the deep, elastic hair turned in; the outermost of sealskin decorated with a simple design. The upper part of his body was covered additionally with a duffel dickey-coat, with pointed hood attached, and in very cold weather he wore an outer pair of trousers. On his tiny hands were shapeless mitts made from ankle-skin of caribou. Fully accoutred, Pete looked like some odd animal: all furry and woolly save where a patch of his brown face peeped through the dog-fur edging of his *kooletah*-hood.

The daylight hours were all too short for Pete. He had to crack his miniature walrus-hide whip and batter the smallest pups until they yelped for mercy. Yesterday's toys he had to dig from the drifts. Circles, triangles, and squares he had to track in the soft snow. Nor could he neglect the morning's inspection of bottles, cans, corks, and broken dishes at the post's garbage-heap.

[1] "I am hungry!"

When the wind was blowing and the snow drifting, Pete and his friends played in the shelter of the house. But when the weather was fine they raced to and fro on the harbor, hauling each other about on small *komatik* or playing tag. Sometimes they spent the entire day building a small, lopsided *igloo*. They were not permitted to use the big *pana*-knife; so they had to dig the snow out as best they could. Occasionally they built up a fat, legless snow-man or snow-bear or snow-walrus.

Pete frequently played by himself. But he never appeared to be lonely. Sometimes he wandered out to the middle of the cove and followed cracks in the ice, placing one foot just in front of the other; or whirled himself about until he was dizzy. He laughed gaily as he fell, or screamed at the slabs of green ice to see whether he could make them hurl back a faint echo. He was deft at somersaulting. He could flop over the precipitous ends of drifts most adroitly, though why he didn't break his neck or collar bones is beyond me. He had a trained corps of bottles and cans lined up on a smooth snow-bank: down these he would roll on belly or buttocks, shouting gaily. He had a favorite block of ice out on the harbor where he sat by the hour playing at "dog-team." In front of him, their feet grotesquely stuck into the snow, was a "team" of frozen fox bodies, skins removed. How unspeakably chilly they looked, their pale flesh exposed to the wind, their teeth grinning oddly! Sometimes they were harnessed with strings or strips of cloth. Whip in hand, urging each "dog" on to the utmost of its strength, Pete would shout "Ou'ah!" and "Howk!"—groaning and chuckling and grunting like the most experienced driver of the *komatik*-team on the island.

Pete's day was not without its excitement. Sometimes he stubbed his toes and fell. Crying loudly, digging his fists into his eyes, he stood stoically while the pain wore itself out. He did not run to his mother. Sometimes he froze his cheeks or chin, or more rarely, his nose. There were moments of real danger, too. Husky dogs, even the most friendly of them, may become excited at seeing a youngster running about, especially if he is wearing pads of bear-skin on his outer boots; they may give chase or actually pounce upon the child if he happens to stumble and fall. The dogs who lived regularly at the post were a docile lot; but there was no telling what the half-wolves from Native Point or Cape Low might do when Pete was playing by himself.

The great event in Pete's day was the evening's visit to the house where Sam and Jack and I lived. After supper, Mary Ell and Ookpik came, in response to a signal from our kerosene lamp, to wash the dishes and sweep the kitchen floor; and Pete usually accompanied them.

He came in shyly and, if not addressed, sat simply at the table "scoffing up" the remains. In these operations he was not hampered by knife, spoon or fork. If we gave him a word, a smile, or a glance of encouragement he climbed upon us, begging to be tossed about in a game called "One, Two, Three!"

Before Pete and I became acquainted he would quiet down immediately upon my coming into the room. Stealing about he would haul a copy of *The Spur* from a pile of magazines, and after ensconcing himself on the floor, his face turned away from me, he would turn the pages silently, examining the pictures. Sometimes, upon finding the photograph of an animal that he recognized, he would whisper its Eskimo name hoarsely. If I spoke to him, he shrank, became sad-faced and speechless, or gave forth a sigh of resignation that vastly amused everyone in the room excepting me. I smiled at Pete's behavior, you understand, along with the rest. But I was not amused; I was a little hurt.

Now that we had learned more about each other, Pete was glad to have me turn the pages of the magazines; and his face lit up when I talked to him, in half-sentences, in his own tongue. After a few minutes with the pictures, acrobatics usually began. I put a record on the gramophone and Pete danced by proxy, as it were, standing on my feet. I tossed him in air, dragged him about by one foot, let him walk up me to "skin the cat," or flung him over my back like a sack of beans. He was very susceptible to tickling. A touch on the chin fairly sent him into fits. His laugh was so compelling that we all laughed with him no matter how we felt.

After a quarter of an hour's romping Pete usually asked to go upstairs to my workroom. As we mounted the steps Pete's mien changed. He was approaching a Holy of Holies: a place where he must make no noise; where he must not touch anything without permission; where he would hear more than once the strange word "care-foo."

He was interested in every blessed object in the room. He asked endless questions, most of them unintelligible. I opened a bottle of carbon tetrachloride for him to sniff: he wrinkled his knob of a nose in an odd grimace. We set and sprang a mouse trap; and he shouted in glee, asking again and again to have me set the trap so he could be the mouse. He crept up to the trap, tickled the trigger with a broom-straw or piece of cotton and waited for the moment of supreme excitement with as much concern as if he were harpooning a polar bear. The tape that *zipped* back into place altogether fascinated him, the more so because he was not allowed to play with it. Pen, ink, books, bottles, needles, wire-cutters—all had to be examined. Drawers had to be pulled out, boxes opened, a match struck, a candle lit, the accordion or the harmonica played. He greatly enjoyed having his fingernails snipped off with a pair of curved scissors.

If left to his own devices, Pete invariably turned to the collection of stuffed lemming skins pinned in rows to sheets of cork. These he examined again and again, touching them very lightly and saying to himself over and over: "*Oonah uvinghuk. Oonah uvinghuk.*" [1] Then he counted them in a businesslike voice. He counted them, I should say, a dozen times each evening, but he never got a larger grand total than *tidlimoot* [2] (though thirty or forty specimens were always on view) and he left out *two, three,* and sometimes *four,* in the process of counting. Sometimes I let him hold a ptarmigan- or hare-skin in his arms. At such moments his face reflected a deep, wondering joy.

Pete could make a good deal of "music" on the accordion without help from anyone. Sometimes he sat on the bed entertaining me while I wrote my diary.

Sooner or later, he was sure to return to the question of the little hatchets. He once found under my table a box containing six small, sharp, bright hatchets, and he lost no time in telling me that he liked them, wanted them, and in fact *needed* them. Each evening I told him that he was not old enough to use a hatchet; and each evening he asked, with a melancholy clouding of his eyes, whether he were not "old enough now!"

Sometimes we entertained Pete in the kitchen, giving him some unusual dainty to eat. Once we bade him open his mouth and shut his eyes while we squeezed lemon juice onto his tongue. He sputtered a good deal, shed a few tears, and burst into laughter, shouting in choking voice, "*Ateeloo! Ateeloo!*" [3]

By eight o'clock, the evening's work was done and Pete had to go back to the servants' house. This was the day's chief

[1] "They are mice! They are mice!"
[2] Five.
[3] "Again! Again!"

moment of sorrow. Always he wanted to stay; but always he had to go. We waved and shouted "Good night, Pete!" as gaily as we could, but Pete did not wave. His face was sad. He buried his head in Mary Ell's *kooletah*-hood and wailed. His wailing was like the wailing of a swarm of great mosquitoes.

Chapter XX

Mikkitoo

You have heard about my young friend Pete. Now let me tell you of Mikkitoo, one of Pete's "wild" relatives.

Mikkitoo did not live at the post. He lived at Native Point. He did not have a wooden floor to play upon in stormy weather. He never ate any left-overs from the White Man's table. He was a fine lad. I liked his broad, cheerful face. I liked to watch him at play. I liked to watch him working at his weasel-snares, at filing down bits of *tooghak*-ivory, at hooking sculpins when the tide was out.

I have thought a good deal about Mikkitoo since leaving old Shugliak. Even as I write he is working or sleeping or playing or eating "way down North" in his barren island-home. He must be ten or eleven years old by now. He has no idea, himself, that he has lived ten or eleven years, and it is doubtful that his elder brothers and sisters, or his parents, or even his grandparents have any very definite concept of his age; for Shugliak is a timeless island.

Mikkitoo is too young, even now, to have thought much. He has his ideas, but they are not involved ideas. He has seen enough of the world to know that some experiences are pleasurable and that some are not. On the whole he finds the long days of summer happy days with their warm air, their pretty moss flowers, and their broods of young ducks; and winter, with its brilliant stars and its talk of wolves and caribou, a time of great excitement. Summer to Mikkitoo is not a short, chilly, fog-hung summer: it is simply summer, season of the singing birds; and winter is not long and dark and dreary: it is simply winter, the time when the ground cracks because it is cold.

And what of Mikkitoo's education? Can he ever become an *angekok*, a leading man of the Aiviliks, without sober hours in a schoolroom? Can he achieve culture without learning to read and write? He is old enough, now, to have his own *pana*-knife, his own bow and arrow, his own bolas. Already the biggest dogs in his father's team respect his word of threat or command. Mikkitoo is becoming an *angot*, a man. He is growing slowly, for the Innuit always grow slowly, but he is growing nevertheless.

Mikkitoo cannot attend any sort of school, for he does not live near the Mission. What he learns he learns from his father and mother and from the other older persons that live at Native Point; or from his own foraging expeditions across the tundra and along the shore.

Mikkitoo has no alphabet to memorize. The language of his people fairly pops and crackles with prefixes and infixes and suffixes; and he will, eventually, have to learn the precise meaning of these. But his ancestors perceived no need for separating their words into the smallest component sound-parts, for giving any such sound-parts special names, or for devising symbols of them that might be painted on skin or graven in stone. When Mikkitoo looks at the White Man's magazines at the post he doubtless realizes, to some extent, that the marks so much in evidence on the paper have significance; but he has no printed literature of his own. His elders do not even write letters to one another unless they happen to have learned the syllabics that have been introduced by the missionaries.

So Mikkitoo spends no time learning how to spell or write

down even such short, simple words as *ikki*.[1] Nevertheless he must learn to speak, and to speak correctly. He must say over the names of the objects he sees about him, thus learning the pronunciation of these nouns. He must master, even during his tender years, some of the well-nigh innumerable inflections of the commoner verbs. His sentences may be short, but they must be cast properly. If he makes a mistake, his elders correct him.

Little by little, as Mikkitoo grows older, he learns more about his tribal tongue. Since he is an Aivilik, he must pronounce *nanook* and *tupek* exactly as an Aivilik pronounces the words, not as a person of the Okomiut would pronounce them; for dialect is important. He must perfect his enunciation; not that he will try to speak more clearly, for good Eskimo is not necessarily spoken clearly; but he must know how to talk deep in his throat; he must be able to pronounce certain words in his mouth without even using his vocal cords; and he must be able to produce, between his cheeks and molar teeth, that peculiar, liquid accompaniment to the purely vocal sounds that makes of the Eskimo language (for us of other races) a practically unwritable language.

As he travels about with his parents he is constantly running into new words. Each day brings him a new sensation, a new color, a new sound, a new mineral, a new cloud formation: all these mean new adjectives and nouns. When first he sees a whale hurling itself from the water, or a lemming nibbling willow-bark, he learns a new verb.

During the shut-in days of winter, Mikkitoo learns the stories of his tribe. In hearing these, he becomes acquainted with the traditional beginnings of the Innuit; in telling them, and eventually in composing tales of his own, he creates his literature. His people have no papyrus, no tablets of clay, no

[1] Cold.

caribou-skin parchment, no stylus, no quill—but they have their Iliad, their Odyssey, their Beowulf all the same. They have no linotype, no printing press of any sort, yet they have their Æsop's Fables. They neither send out nor receive announcements concerning the latest books, but the chances are they have even their own *Story of Philosophy* and *Mind in the Making*. This literature with which Mikkitoo becomes familiar, and to which he will eventually contribute, is not, precisely speaking, a literature at all, for it is composed not of *letters*, but of spoken words. But it is, to the Eskimo, a definite form of culture, and it is rich and varied. Furthermore it is probably quite as satisfactory to young Mikkitoo and his comrades as our English literature is to you and me.

Mikkitoo has no library, but he has wondrous discussions to hear in the dim light of the *koodilik*-lamp. He has no favorite anthologies, but he will memorize, in time, innumerable songs and poems: love ballads; poems about bears, weasels, caribou, and whales; narrative songs; introspective poems; humorous "jingles"; satirical poems; even epics.

And he will learn many tales, about all sorts of persons, mythical beings, and animals. He probably never will memorize any long essay on *Contemplation*, but he will be able to tell a story about a man who thinks a great deal. He will never deliver tedious dissertations upon world-trends, but he will tell dashing stories about what happens to persons and peoples who behave in a certain manner. Many of his tales will be comical for their very brevity. Others will be so long that hours, even days, will be required for a really adequate telling. Mikkitoo cannot, as a child, tell these stories save to his younger companions. But one day, when he can speak perfectly, and when he has memorized them well, he will tell them, with ceremony, to his family and their friends.

Not all Eskimo men are equally good as story-tellers. Not

many of them become leaders of the tribe. But Mikkitoo, even if he never becomes an *angekok*, will compose songs and tales of his own. Some day he will kill his first *nanook* and there will be a new song for the tribe to sing. Some day, at the floe, he will have a narrow escape from death, and the children in the *tupek* will listen open-mouthed and wide-eyed to the story of Mikkitoo and the Great Walrus.

Mikkitoo has little usè for Arithmetic. He may learn to count to ten, using a pile of pebbles, or pointing to the fingers of his hand; but the chances are no one will bother to tell him about figures higher than these. He has no wall-paper problems to solve, no square roots to extract, no fractions to divide. It is doubtful that he will ever have a very definite concept of such a large number as one hundred. If there are more than ten or twelve walrus in a herd at the floe, there are simply *aiviuk amishueet* (many walrus) or *aiviuk amishualueet* (a great many walrus) there. In Mikkitoo's matter-of-fact world no one cares to know that one hundred is ten times ten so long as there is enough blubber in the *koodilik*-lamp, or enough hide to make good harpoon-lines.

Mikkitoo's education is direct. The questions he asks are answered by his elders. He may ask the same question again and again, but the answer is always given with dignity. Mikkitoo respects older persons. He obeys promptly. Only rarely is there occasion to slap or even to scold him. No one ever beats or whips him. If he misbehaves someone says: "The Innuit do not behave as you are behaving."

Mikkitoo reveres his dead ancestors. He knows that the spirits of these relatives are reincarnated in himself and in his brothers and sisters. So firmly does he believe this, that if you ask him to tell you his brother's name he may answer, "That person is my grandmother."

Mikkitoo hears much about the spirit world. He listens to

the edges of the skin-tent flapping in the wind and learns that the spirits are abroad, trying to find a warm place by the *koodilik*-lamp. He sees fog drifting in from the ocean, and learns that the spirits are sad, and have hidden themselves behind the white vapors. He watches the aurora borealis flaring and shimmering across the sky, and learns that the spirits are playing a game with a man's skull. Without even remotely realizing it, Mikkitoo becomes an animist. Every rock, every willow-shrub, every sea-gull, every weasel, every cloud, every sculpin in his world becomes, for him, the dwelling-place of some being. Small wonder, perhaps, that he so rarely finds himself lonely; for he really believes that spirits are living in the objects all about him.

Mikkitoo is not a grind. He learns to pronounce his words correctly; he thinks now and then about the spirits; he has vague ideas about being a worthy man in his tribe. But his daily, his almost hourly interest, centers in playing, in learning to drive the dog-team, and in hunting. He longs to grow up so that he may go to the floe to shoot seals and walrus, to the highlands to hunt caribou, and to the rocky islands to chase *nanook* with dogs.

It is as instinctive for Mikkitoo to chase, capture, and kill small mammals and birds as it is for a Husky pup to dig in the moss for lemmings. Not often does he make pets of wild creatures. If he catches a weasel he skins it carefully, stretching the pelt out as best as he can, or stuffing it crudely with moss. He learns to set snares. Sometimes he catches shore-birds with nooses laid in the sand. He makes little nets for trout. But as he skins the soft body of a lemming he is dreaming of fierce encounters with bears; and as he sets his nets in the shallow pools his thoughts are of huge, black whales.

Mikkitoo is too young now to know anything about love. He is fond of his parents, for they give him food when he is hungry, they tell him wonderful tales when it is stormy weather, and help him to set snares. He is fond of his sisters and brothers and of other children his own age, for they are all good playfellows. But he has no sweetheart. He has heard many times of the girl the tribe have chosen for his wife—a girl who lives on the mainland, two hundred miles away, the daughter of a famous Aivilik hunter. He has sent this girl presents as formal tokens of his esteem. He knows her name. But he does not love her at all. He has never even seen her. Most of the time he is not in the least interested in her.

Mikkitoo knows there is such a thing as sex in the world. Has he not been told he is not to go into Tooahtak's *tupek* for many days because Tooahtak's wife has given birth to a baby? Has he not marvelled at the perfectly formed young *netchek* the men cut from the mother-seals in the season called *Netchialoot?* Has he not laughed at the funny strings of embryos the trappers pull from the bodies of female foxes? Has he not heard long discussions about the male hares that give birth to the young? [1] Has he not listened to the mournful cry of the slim bitch-leader of his father's team as her big mate is whipped and beaten and taken away from her? All this, and more, has Mikkitoo seen and heard. Yet not one word has the great god Eros whispered directly into Mikkitoo's ear; for Mikkitoo is very young.

A few years more and Mikkitoo will begin to understand the dancing of the long-legged cranes, the hollow booming of the owls, the unceasing spring flight-songs of little *Kungnitook*,[2] the longspur. A few years more and part of his heart will go out to the slim bitch-leader as she mourns for her mate. A few years more and summer for Mikkitoo will be far more than the season of singing birds. It will be a season

---

[1] There is a widespread belief in the North Country that the Arctic hare is hermaphroditic.

[2] Lapland Longspur, *Calcarius lapponicus.*

that he cannot name; a season so bewildering that he will walk off across the bog-cotton marshes by himself, breathing in the odor of the moss-flowers, listening to the low sounds of brooks and bumblebees, watching the clouds in the wide sky, trying to find somewhere, among all these happy spirits, words beautiful enough to express in some permanent way this new feeling, this new interest in life, this new awareness that has come over him. For Mikkitoo will be in love. Perhaps with the black-eyed daughter of the famous hunter at Repulse Bay; perhaps with one of his boyhood playmates; perhaps even with a pretty girl of the Okomiut. And Eros, deigning at last to recognize our Mikkitoo, will talk with him on the tundra.

Mikkitoo may not marry for many seasons. His family may have to cross Sir Thomas Roe's Welcome by motorboat in order to have the wedding ceremony in the traditional land of the Aiviliks. He may fall in love so fiercely with a girl of his own choosing that the tribe will be obliged to change their plans. He may even leave the tribe, if need be, in order to live with the girl that he loves.

Mikkitoo will love his wife and care for her day after day. He will love his children. He will beg the soothsayers to tell him that he will have many boy babies, boys who will grow up to be hunters of the tribe. He will hunt caribou in the Fox Channel highlands so that his family may have proper clothes for winter; he will catch many foxes so they may have slap-jacks and tea, duffel *kooletah*, graniteware mugs, and perhaps an accordion. He will expect his wife to help him howsoever she can, in sewing, in caring for the *nootarak*,[1] in scraping down skins, in cooking if cooking is to be done, and in chewing his *komik* until they are wearable. But Mikkitoo

[1] Babies.

will do all the dog-driving, all the hunting, all the trapping, and all the *tooghak*-carving.

And when it is the winter season Mikkitoo will gather his family and their friends about him in his own *tupek* at Native Point and tell them the old, old story of Ookpikjuak, the Great White Owl, and the two silly hares that let themselves be caught while they were making love.[1]

[1] See Chapter XXXVII.

CHAPTER XXI

My First *Igloo*

You have a fair idea, by this time, of life at the trading post. Sam and Jack and I were having a grand time together, they with their trading and "office-work," I with my scientific studies and painting. We played games every evening. Most of the time we were too busy to read, though we liked to peruse the pictures in our store of magazines. Occasionally the Fathers from the Mission came in for hours of Poker with white beans for chips; or for long, fierce sessions of Hearts. Bridge was too much for us, and Sam called Whist "an old woman's game." We frequently made candy: fudge poured out over a plateful of raisins or currants.

But I wished to see more of the island and to learn more about the Eskimos. Now that I could say *"Ahnoway amishualueet oobloomi!"* [1] without appearing guilty of some indefinite crime, I decided I was ready for a caribou-hunt with John Ell. My face was sore from freezing, but I was impatient. We decided to go to East Bay along the high, eastern shore of the island, about a two days' *komatik*-journey from the post. We would leave on November 22, if we could. We would meet Muckik *en route*, one day's journey eastward.

The morning of the 22nd was cold and clear, a perfect morning. We set off at about ten o'clock. Through frost-spangled sunlight we sped across the rose-white plains, now rattling over a broad lake so shallow it had frozen to the bot-

1 "Very strong wind today!"

tom weeks before; now skirting smooth-topped plateaus of gravel where, in the wind-swept areas, our runners grated noisily over sharp-edged clam shells—remnants of a distant age when Shugliak was part of the ocean's floor. We stopped for lunch in the middle of a big lake. John made a windbreak of snow-blocks, set the Primus stove [1] to burning, put a kettle of snow on to melt, and turned the *komatik* upside down. We drank some tea from a thermos bottle; and chewed up some hard biscuits, a great quantity of buff-colored cheese that crumbled in the cold air, and some raw frozen fish. We glazed the *komatik*-runners before travelling on.[2]

By one o'clock in the afternoon the sun was so low and the western sky so heavy with smoky clouds that we decided not to go farther and, should the weather permit, to resume our journey in the morning.

John turned the dogs sharply to the left, heading them for a low ridge where, it was explained to me, the drift was deep enough for making an *igloo*. Fully an hour passed before we reached the ridge. As the glow of sunset subsided our eyes became accustomed to the twilight and we could see our pale surroundings distinctly, even though clouds were mounting rapidly and spreading over the sky.

John, who looked like a strange piece of overstuffed furniture in his thick winter clothing, said little. Having reached a drift apparently to his liking, he passively stopped the dogs and, leaping from the sledge, bid them lie down. They had not had a hard day and were in frolicsome mood.

[1] A small kerosene stove for Arctic travellers, manufactured in Sweden.

[2] *Komatik*-runners nowadays are made of wood. They are coated with an inch-thick layer of what Sam called "mud," a mixture of humus and clay, patted on evenly, frozen hard, of course, then glazed with water or urine. The Eskimos are careful not to chip the "mud" from the runners, and when they are travelling they stop several times a day, if they can, to put on a new ice-glazing. This makes hauling much easier for the dogs.

With the cleaning-rod of his rifle John proceeded to probe in the snow for a drift of proper depth and consistency. It was important, I learned, to find a deep drift that had formed during one storm, for blocks cut from many-layered drifts may break into two or three pieces. With his long *pana*-knife he cut a neat slab about three feet long, eighteen inches high, and eight inches thick, digging the snow away from one side with his knife and his hands, and kicking the block loose with his foot. Having thus made an opening in the drift he hewed other blocks out more easily. Six blocks soon lined the trench in which he stood.

Cutting one several inches longer than the others, he stood it on its longest narrow side, bridgelike over the trench. Against its end he set another block, the top of which he cut down a little. Soon he had erected, end to end, an ample circle of blocks, the first one low, sloping up from drift-level, the last of full eighteen-inch height. He worked steadily, lifting out the solid chunks and placing them skilfully end to end. Soon the low dome began spirally to rise, the wall now sloping inward strongly, but kept from falling by the strength of the packed snow at the seams, which hardened after a moment's exposure to the cold air. John worked altogether from inside, quarrying his blocks from beneath the rising structure. The lower third of the *igloo's* wall was the solid drift itself. The entrance was the opening under the first block he had set in place.

I was so deeply interested in the construction of this, my first snow-house, that I scarcely realized how dark it had become. The sky was overcast; light snow was falling. The drifts about us were dull blue and their outlines dim. I clomped about, now and then swinging my arms or beating my hands together to keep the blood racing. The dogs had

not whined nor moved since they had curled up, fully harnessed, in the snow.

Suddenly John stopped his work, and stood silent. Three
dogs rose, as at a mysterious command. Had I, too, heard a
sound? The rising wind lisped along the edges of the half-
built snow-house. More dogs rose; one of them uttered a
low growl.

Out of the gray night that hung so heavily about us came,
clearly now, the whine of a dog. Our team started to bolt,
but, ordered to keep quiet, stood stiff-legged and tense. Soon
we heard the patter of padded feet, and the scraping of
komatik-runners. From the westward, his dogs following our
trail, had come one of the Eskimos from Munnimunnek
Point. I could not see the newcomer's features in the darkness, but I knew from John's tone of voice that his coming
was a surprise. He was, I learned, to spend the night with us.

An air of mystery surrounded this unexpected arrival.
Why had he joined us? Why had he not started to make his
own snow-house at the fall of evening? His voice was exceedingly rough, and his dogs, who now fought fiercely,
were, I perceived, half-starved.

John told me that our visitor was Khagak. Khagak shook
my hand heartily, as if he were an old friend. Why hadn't
I remembered the man? Or had I, indeed, ever really seen
him before? His laughter was so rough and his talk so in-
comprehensible that I longed to see his face.

John bid Khagak help him with the snow-house. Roaring
an assent, Khagak crawled inside and carved out blocks while
John put them in place. Steadily the dome rose. Finally
came the greatest test of the workman's skill—the fitting in of
the topmost, key block. John crawled outside for a moment
and placed a block on the unfinished roof. Returning to the

inside, he stuck his pudgy arm out of the opening, pulled the block toward him, balanced it traylike on his hand and, whittling it down to the proper size and shape with the *pana* held in his other hand, he settled and chinked it down into its important position. John was an expert builder of the *igloo*. The walls, roof, and floor of our house had been finished within an hour.

To keep myself warm, and also hoping to be useful, I had busied myself with packing wedge-shaped pieces and loose snow into the cracks, walking round and round in my search for crevices among the thirty-nine blocks that had been used in the *igloo's* making. I was heartily thankful when John said that it was time for me to come in.

Stooping to hands and knees, I stuck my head and shoulders through the low opening that served as the door, and wriggled forward. My shoulders, broader than John's or Khagak's, were further enlarged by many layers of clothing. Eager to leave that cold outer world, I shoved hard, lifted my body a little, and crash! down came a block. I was afraid that in one false move I had wrecked an hour's labor. The Eskimos laughed, however, and I knew the damage was slight.

Inside, the darkness was oppressive. I could barely stand in the structure's very centre, and if I moved a little in any direction my head or shoulders precipitated a shower of snow. I sat down on what felt like a snow-ledge while John went out after a candle. Somewhere near me was that rough-voiced Khagak. As my eyes adjusted themselves I made out dim light where the wall was thinnest, or where my patchwork had been insufficient. The door darkened, there was a muffled scraping, and we heard John fumbling in his clothing. With a snapping click he struck a match on his thumb nail. The brilliant flare momentarily blinded me. As the cord wick of the candle caught the blaze, an even tongue reached up, wavered a trifle, and burned steadily.

John smiled at my gasp of delight. With the darkness vanished the oppressiveness that had been so noticeable as I entered. The symmetrical, smooth-walled dome, about ten feet in diameter at its base, shone with an almost supernal radiance, the immaculate blocks reflecting and diffusing the candle's small fire so perfectly that we were in an all but shadowless world. The candle was set in the snow. Tiny particles of rime hung in the cold air about it. These floated slowly and gracefully downward, to the floor.

Chapter XXII

Three Men in a Snow-House

THERE I sat, in my first *igloo*. I was an Eskimo at last. It was all quite wonderful.

For a moment I had completely forgotten Khagak. My eyes now fell upon him as, crouched on the floor not three feet away, he watched me silently. He was no ordinary Eskimo. His coarse, black hair was long and stringy, all but hiding his low forehead and his small, red-rimmed eyes. His nose was blunt and wide; his mouth was covered with a thin, shaggy moustache; and his cheeks, blackened by repeated freezing, were cracked and scabbed. His clothes were of the usual caribouskin, but they were exceedingly old. Suddenly he half rose, smiled broadly, and shuffled out to free his dogs. I could hear him tossing the coiled harnesses onto the *igloo's* roof.

While the hubbub continued outside, John told me that I had seen Khagak before—that he was the man who liked to play the accordion. Instantly I remembered him. He was the notorious Khagak, *The Hungry, The Cannibal.* I remembered the report that years ago, when he was at the point of starvation, Khagak had shot his little daughter in the chest, ordering her to bring water in which she was to be boiled. The bleeding child had been able to crawl to a pond and to return with water before she died. And Khagak had cooked and eaten her, even as he had threatened. As I thought of Khagak and recalled this hideous story, I marvelled that his laughter could be so boyish and so frank. And I decided that the tale must have been told either to amuse or to horrify me; that if Khagak had ever harmed anyone it was only because he had been utterly out of his head with hunger.[1]

The ledge of snow on which I sat was about one and one-half feet high and occupied more than half the *igloo's* floor-space. On this ledge the sleeping bags were soon stretched out on bear- and caribou-skins. Khagak set up his stone *koodilik*-lamp. John lighted the Coleman [2] and piled snow in a kettle. There was little for me to do save watch with keen interest the movements of my companions. I was very hungry, for our breakfast at the post had been small, and our noon meal almost worse than nothing. I expected, at any moment, the tantalizing odor of bacon frying, or of canned meat simmering. But no such odor rose. Tea was in the making. John pulled from a sack a two-foot *ichalook*-trout and with a hatchet chopped a neat slab from its back. This he offered to me.

"Do you suppose I can eat it raw?" I wanted to ask.

"*Pitchiak!*" [3] John said, apparently reading my mind.

My question would have been somewhat rhetorical; for I knew perfectly well that I could eat it raw. I had eaten raw fish at noon. But the prospect of more frozen *ichalook* as a "square" meal somehow confused me. For a dreadful instant I was speechless, motionless. I remembered, in that instant of hesitation, how I had dreamed of the romantic life the Eskimo leads in his cozy *igloo*, how I had resolved to get on with this people, to learn their ways, to become so much a part of their lives that they would not be afraid of me. All

---

[1] The killing of daughters, especially at times of extreme food-shortage, is not unheard of among certain tribes of the Innuit, even today. To the best of my knowledge, the Aiviliks have never found it necessary to resort to such drastic methods of tribe-preservation. As to the Okomiut, I cannot say.

[2] Gasoline stove.

[3] "Good!"

these reflections while a slab of frozen fish was being proffered on a damp brown palm. An instant's hesitation—and I took the fish.

The icy chunk chilled my fingers. With a grim resolve to keep down what I swallowed, I put it to my lips. As I chewed, the hard, orange-pink plates of muscle thawed and fell apart. The gray skin became slippery. I tried to forget the indescribably repellent odor. I swallowed a mouthful, and could feel the coldness drifting down to my stomach. Mouthful after mouthful went down.

I now sank my teeth into the slab, determined to get something to eat. Heedless of skin and bones I chewed savagely, though I was already quaking and my teeth throbbing with the cold. I think it would not have tasted so bad had not the backless *ichalook* been lying there in front of me, mouth agape, frozen eyes staring, with just the expression of one I had sketched in water colors, not long before, at the post.

John and Khagak, meanwhile, had been consuming quantities of fish. John now turned round. "More?" he enquired politely.

"I'm not very hungry," I answered. "But I'll have some tea."

He handed me a granite mug. Thankful for its steaming hotness, I put both hands about the mug and gulped down the tea. Then I fell upon the store of ship's biscuits and crunched till my jaws ached.

As John was flinging tea leaves about our spotless ceiling and putting aside the mugs, he explained that I would never starve so long as I was with the Aiviliks. I was glad I had eaten the fish though the thought of a repetition of the meal almost made me sick. I felt better after I had kicked aside that face that kept leering from the floor.

As the *igloo* became warmer I was more comfortable save

for the odd, somewhat unpleasant odor that became steadily stronger. This odor was not to be traced to any single source. It was damp clothing, old fish, gasoline, dogs, seal-oil, fox-bait—all these and more combined, and it emanated from everywhere. As the heat increased, Khagak began coughing, —a deep, wet, racking cough that made me shudder.

Presently Khagak rose from his place near me, dragged into the open a huge chunk of *kellilughak*, and began to chop up dog-food. John seized his axe and pulled several decayed fish from the bag that had held the principal item of our evening meal. How he recognized the present from the perfect and pluperfect tenses in these frozen *ichalook* is to this day a mystery to me. The walls resounded with the whack of falling blades. Full two bushels of "muck" [1] were chopped up—two days' rations for the dogs. At the first sound of axes, the dogs began to whine and to run excitedly round the *igloo*, sniffing loudly at the cracks and putting their feet on the roof. Now and then John shouted out some extravagant threat such as: "Keep away, or we will all put our feet on your stomachs and pull your tails off!"

The chopping finished, John crawled outside, beat off the dogs, and waited for Khagak to push the meat out with a piece of antler. The dogs were beyond control. They rushed to the feast like water through a breaking dam. John struck one and it howled savagely but forgot its pain the moment its nose led it to a chunk of flesh. The meal was consumed in about five minutes. The chunks were gulped whole. The fighting was furious, since our dogs did not know Khagak's, and the newcomers were very hungry.

When John came in, he pounded his clothing with his snow-beater [2] and sealed the "door" with loose snow. As we

---

[1] *Muck* is North Country slang for *food*. It is not a synonym or abbreviation of *muckluck*.

[2] A specially constructed, paddlelike piece of wood.

sat on our sleeping-bags we had little to say. In the moisture-laden atmosphere the candle burned dimly. The Eskimos smoked endlessly; my eyes grew weary of peering through the haze.

"Where are we tonight?" I asked, supposing that John would tell me how far we were from the post.

To my surprise he replied, "*Kokshowktok.*"

"Why do you call this place 'Kokshowktok'?"

SAD-VOICED *KUDLOOLIK:* THE PACIFIC LOON

"Many, many much loon.[1] Some many, much far; some not so many, much far; some mebbe not so far; some much far; some not far; some up, mebbe close, mebbe far; many, many loon, all *wahoo! quack! wahoo!*" John's *Kokshowktok* was a veritable universe of loons.

John had, you see, learned a good deal of English. He told me, in fact, that he knew three languages aside from his own—English, Scotch, and American.

I now reminded John of the fox he had caught that day and which he had said he would skin that evening. He

[1] *Kokshowktok* means *A place where loons live.*

reached for the bag that had held our evening meal and drew forth the carcass. He had put it in the warmest place he could find to keep it from freezing, but two of its feet and the tip of its tail were stiff. He now sat on the tail and held one paw while I held the other until they were thawed out. Then, with his *ikpiakshuk*-knife [1] he cut the skin from one hind foot directly across the body to the other hind foot, and pulled the fox inside out. The long tail, its many vertebræ joined with strong sinews, came out all at once at a deft jerk of his powerful fingers.

As John skinned, I remembered the pretty creature as it had leaped about, held fast in the trap. Now I stroked its round ears, tested the sharpness of its claws, and admired the perfection of its immaculate coat. John rolled the pelt up, tail and head inside, and, after tying it roughly with two of the limp legs, tossed it back into that most convenient, all-welcoming bag. I had witnessed the first lap of *Teregeneuk's* journey to the smart Outside World.

[1] Pocket-knife.

CHAPTER XXIII

Nocturne

IT WAS scarcely seven o'clock; but the trip in the open had made me drowsy and I was vaguely cold, even in the shelter of the *igloo*. Conversation with my friends was difficult. I decided to go to bed.

The sleeping-bag into which I wriggled, feet first, was a trifle short for me, but it was adequate. The inner bag was made from three winter caribou-skins, the deep, amazingly warm, and elastic hair turned inside; the outer bag, just large enough to encase the other, was of handsome *kashigiak*-seal, the hairy side out. I needed no other covering, though I found that the opening at the head let in cold air unless I stopped it with some of my clothing. Though I did not altogether enjoy the odor of the atmosphere, I was more comfortable than I had been for many long hours.

Soon the Eskimos took off most of their clothes [1] and went to bed. Lying on their stomachs, heads supported by elbows, they smoked their pipes. John blew out his candle; Khagak's *koodilik*-lamp burned on, however, a filament of black smoke rising from the orange flame to coil among the pendants that gleamed from the ceiling.

John, like all good Eskimos, went promptly to sleep. Khagak, however, was taken with a fit of violent coughing. He squirmed and turned, and finally rose on his elbows to

---

[1] It is commonly believed by "civilized" persons that Eskimos never take off their clothes. The Eskimos I knew at Shugliak undressed practically every night.

smoke his pipe again. His coughing continued. Now and then he made brief remarks to himself. The significance of these remarks I did not catch; but the commotion kept me from sleep. In fact, as I lay there, increasingly conscious of all that had taken place that day, I felt less and less drowsy. I may even have been nervous, for I found myself listening, neck muscles taut, for some new note in Khagak's coughing, for the growl of a dog, or for the low sounds of lemmings running through their burrows not far from my head.

Presently there was a moment's silence, startling in its intensity. The wind had abated. No dog stirred. John's breathing was so quiet that it seemed he must be lying awake, listening as attentively as I. At last Khagak had fallen asleep and the world, for the briefest of intervals, was blissfully at rest. The *koodilik*-lamp's light was steady and friendly. From it came warmth and healing. The cold and wind were far, far away. I turned over quietly, ready to sink to sleep.

To sleep? Rising from the depths his cough had penetrated, like a mingled roar and growl, came Khagak's snore. So hideous was the sound that I half rose, in alarm, to gaze at the man. At every inhalation his moustache twitched; as the breath came out in a snarl, his nostrils quivered. He had no pillow, and his face was turned upward so far that I could see his eyes were not altogether closed. I was thankful that the coughing had ceased; but the snoring was so loud that even the dogs were disturbed. As I lay back with a sigh, John also began to snore, but his was the soothing obbligato of a violoncello.

I needed sleep, but realized that fretting over the situation would not help matters any. Contemplating the holes that were by this time appearing in the roof, I amused myself with the idea that while our *igloo* was silently melting away, everything on our floor was frozen stiff. I could see through the crisscrossed dog-harness above me that the sky had cleared. There was a brilliant aurora. The stars were so bright and so low they made me think of the Great White Way.

Khagak turned over, wakened with a snort, and began to cough. He half rose, adjusted the moss-wick of his lamp, reached for his pipe, then abruptly addressed John about some trivial matter, as if such disturbance were a matter of course. John awoke as gracefully as he had gone to sleep, made a quiet answer and began to smoke. I gathered that Khagak had proposed some hunting yarn; he did not cough much when he became interested in a story.

I said nothing. I think John told Khagak that white men make no noise while they sleep. This comment roused me almost to the point of telling him that he had never lived at a college fraternity house nor slept in a barracks; but I sanely held my peace, realizing that tundra vocabulary would fail when it came to "college" and "fraternity" and "barracks," not to mention most of the other words. Finally, after each had told a story, John dropped to sleep, and Khagak, after a spasm of coughing, began once more to snore.

I turned over—but was not in the least sleepy. Through a crack in the wall gleamed a blue point of light, ever so tiny, but bright and crystal-clear. I moved a little so as to note the adjacent stars. Vega! Blue Vega, queen of her simple constellation! So low-hanging was she that if I did not move I could make myself believe her and the dark patch of sky about her part of the queer mosaic of the *igloo's* wall.

As the slow hours passed, I made plans, smiled grimly at man's strange impotence in such a plight as this, marvelled that such a genius as Edison should work so efficiently yet sleep so little, chided myself for not being more constructive in my thinking, and indulged in several fits of frank hatred of the racket my companion to the left was making. Several times I was at the point of raising just as loud a noise as possible in imitation, merely to see what effect this would produce.

I looked at my wrist-watch. In the darkness of the sleeping-bag its radium-plated numerals glowed gaily. A wave of warmth swept over me as I remembered the friends "back home" who had given it to me. They had called it "Sunny Jim"; and "Sunny Jim" had been my constant companion, never failing, never arguing, for three busy months. Even now it went about its business making no useless noise while that Khagak— Good Heavens! could Khagak really be asleep, or was he growling and snarling, snorting and puffing, moaning and snivelling only to entertain me in his primitive way through the long hours? It was almost six o'clock—three hours until daylight. The *koodilik* was burning brightly, though I could see that its supply of oil was low. Khagak stirred, stopped snoring, and after a wolfish cough, rose to put more fat in the lamp.

He did not return to his bed after the seal-oil had been replenished. He drew his *kooletah* over his naked shoulders and pulled on his boots. He uncovered a frozen trout, hacked it in two lengthwise, and began a noisy feeding. It disappeared rapidly: all of it, so far as I could see. Another fish, a two-foot beauty, was hauled out and consumed. Khagak did not eat the bones, I suppose, though he wasted no gesture in seeking or discarding them. He ate some *netchek*-blubber, too, and a hunk of *muckluck*.

John did not join in this early breakfast, though he doubtless would have eaten had he been awake. When Khagak had gorged himself to the point of apparent satisfaction, he scraped his palms with his great knife, ran the blade nimbly about the edges of all his fingers, and licked it off, on both sides, with a loud smacking of his lips. I wondered that he did not lick his hands directly. From a greasy bag he drew forth a gull-skin and wiped his hands—not to clean them, I reasoned, for such a measure would hardly have been economical—but to dry them so they would give him no physical discomfort.

Drowsily I watched him pulling off his *kooletah*. He was, I realized, going to bed once more, to begin his coughing or his snoring.

Now the lights were fading again, the sounds receding. Gradually all the *igloo* became hazy. A deep tranquillity was brooding this perfect world. Here there was no noise, no strife, no thinking, even—for there was no need of thinking . . .

When Khagak returned to his place by me he had somehow taken on gigantic size, and his body was covered sparsely with long, shaggy hair. Fangs glittered between his parted lips. He approached stealthily, first stooping, then hobbling like a hyena, fingers curved like claws. I tried to rise, to shout for help, to burrow into the snow, to find a gun, a stick, a stone—but he was upon me. He put one great knee upon my neck, and another upon my legs, and with a savage growl began pulling out my ribs one by one, licking and munching them with fiendish deliberation.

I wakened with a jump, convinced that I had actually cried aloud. Khagak was snoring loudly and blissfully. He had rolled over so far that his sleeping-bag had all but buried me.

I looked again at my watch—it was now almost seven o'clock. The aurora had vanished. Vega had vanished. In a short time we would rise to prepare for our journey farther into the caribou country.

CHAPTER XXIV

Interlude

I EXPECTED to be weary after that virtually sleepless night, but I wasn't. We had breakfast, wiped our mugs and spoons on a towel, rolled up our sleeping-bags and "mattresses," and crawled to the outer world.

It was a knife-edge morning, the air so cold I gasped. There was enough wind to keep the loose snow chattering over the hardened drift-crests. John and I loaded our *komatik* while Khagak loaded his, and we headed the dogs eastward.

Khagak soon turned south. We waved good-bye. He was going to Native Point. John had given him a little square of paper, a note in syllabics, for Muckik. Muckik, you remember, was to join us on this caribou-hunt, and he had not yet made his appearance.

Khagak's *komatik* was in sight a long while; but it made better time than ours for it was not heavily loaded. We had trouble getting ahead. The snow was deep in the meadow-country and the dogs sank in to their bellies wherever there were willows. John and I had little chance for riding. We had to run alongside the sledge, helping the dogs as best we could. Going was easier on the lakes. Several times we stopped to glaze the runners and to have a mug-up [1] by way of warming our insides.

We saw not a single bird, nor hare, nor fox, nor lemming.

[1] Mug of hot tea.

We noted some fox trails, a few ptarmigan tracks, and the droppings and regurgitated pellets of a snowy owl.

We must have made about ten miles when it began to grow dark. Once more John turned the dogs to the north, heading for shelter. There was another hour's journey. The dogs stopped gladly, for they had had a hard day. John began the *igloo*. I unloaded the *komatik*.

Scarcely had we begun to make camp when we spied Muckik coming: just a sharp black spot on the horizon, a spot that wavered ever so slightly and became a trifle larger as the moments passed.

We were glad to see Muckik, for our party was now complete: three men, two good sledges, two dependable dog-teams. We were a noisy crowd, singing and laughing and growling and *gwuff*ing, working at the snow-house, pounding the bedding, racing round to sniff at one another, all of us thinking more or less definitely about the thrills and joys of caribou-hunting.

Muckik decided he would build an *igloo* of his own: one that should adjoin ours. Boisterously the two Eskimos carved out their snow-blocks, making a sort of contest out of it all, joking and chatting as they worked. The dogs did not fight. Muckik's *igloo* was so close to John's that the walls of the two structures actually overlapped. I wondered at this until I saw Muckik go inside the larger house and cut away the wall. A two-roomed snow-house!

You may think it strange that I did not assist in building these *igloo*. Perhaps I should have. I might have carved out blocks; I might even have set some of them in place. But the chances are I was just as helpful in happily doing nothing as I should have been cracking toe-nails in kicking out building material, or wearing holes in mitts trying to use the *pana*-knife.

The duplex *igloo* was finished in a little over an hour. We moved in. The wind was rising but we were cozy.

Ours was a beautiful abode. In John's and my compartment there was no candle, but the *koodilik*-lamp at Muckik's bedside burned brightly, and there were galloping shadows; a shifting of opal-glow along the archway between the two rooms; a green and blue and orange sparkle of eyes, of finger nails, of teeth, of mugs and knife-blades in our little world of diffused and reflected lights. How could I write a word when I simply wanted to look?

Strange how time passes in an *igloo!* You get yourself comfortable, changing your duffel socks, loosening the collar of your shirt, untying the cord that holds up your caribouskin breeches. You find your ditty-bag, dig out some scissors and a comb and a little mirror, and give yourself a going over. You pass round your lavender-scented soap—a whiff or two for everybody. You find a pack of chewing gum. After you've chewed an hour or so and are ready to throw your wad away you learn that the Eskimos swallowed theirs long ago.

Rifles must be thoroughly cleaned: unscrewed and tapped and pulled and punched apart, the pieces lying hither and yon on boxes, on stoves, on bed-rolls, on the floor. If a piece is lost, so much the better. Now it is necessary to hunt everywhere, turning everything upside down. Everybody's personal effects become so mixed that a complete rearrangement and sorting out is necessary. When the missing piece has been found and tranquillity restored, an hour or two have passed.

Now you inspect your clothing, mending torn places, sewing a little strip of handkerchief along the collar-edge that wears holes in the skin of your neck, hanging mitts and socks up on the antler clothes-rack that is stuck in the wall, and unpacking and flexing the flat, dry *komik* you intend to wear on the morrow.

You get out your diary book at last. "Shall I write something literary," you ask yourself, "something descriptive of this beautiful house?" And then you put down: "Saw a fox track. No birds, only snowy owl droppings and pellets. Hard going through the soft snow." And you look up to see that these companions of yours, these lovable, these altogether unaccountable companions of yours, have also got out their diary books (little books Sam gave them especially for this trip) and they are putting down their ideas in heavy marks that look like higher mathematics of some sort.

The dogs must be fed. You whistle while you chop up rotten *kellilughak* and *netchek*. And you whistle not to dispel the loneliness that has got you down; not to help you forget the awful odor; you whistle because you're happy! Even if your nose is sore. Even if your big toes are throbbing from those too-short *komik*. Even if you can't say to your companions what you'd really like to say.

After a while, you undress for bed. Most deliberate you are, pensive almost, as you skin off your shirt and wriggle out of your breeches, and slip into your sleeping-bag. It is so comfortable there, looking out at the steady *koodilik*-lamp, your face against the caribou hair that is soft, almost, as velvet, only not dusty and voluptuous and sensuous as velvet is, but vitalizing, wholesome, buoyant, with an odor as gently aromatic and as soothing as the odor of pine needles.

CHAPTER XXV

Four Thumps on an *Igloo*

MUCKIK and John were singing when I wakened next morning. I did not recognize the tune, but I noted that certain words and phrases were repeated many times. I wanted to ask what the song was about. Was it an incantation to the gods of the caribou-hunt? Was it a love ballad? But I voiced no question. I simply lay there marvelling at the odd rhythm of the tune, and at the silent dancing of the shadows. It was nine o'clock. The sun had not yet risen.

Breakfast was in the making. The Primus was roaring, coffee was boiling, and slap-jacks were frying. These slap-jacks (*slap*-jacks, not *flap*-jacks, mind you) were great, inch-thick discs of dough, broad as the pan in which they were cooked. We ate a dozen of them.

Soon we were on the tundra again. The snow was firm, the wind low, the lakes numerous. We made good time. Now and then we came upon a landmark: a windswept heap of boulders; a small stone set upon a larger stone; an old antler stuck into the snow. My companions set several traps as we went along, most of them on mounds near the small ponds.

At noon, after a mug-up, we had a shooting contest. My

Krag was just enough bigger than either John's or Muckik's rifle to rouse their curiosity and admiration. The Eskimos shot better than I did. Some of my trouble probably was traceable to unfamiliar thickness of clothing at the shoulders and to half-frozen fingers. After the shooting we had another contest. Each man was to find how many times he could whirl completely round on the ice, standing on one foot. We laughed gaily at our play; but the couchant dogs wore a bored and tolerant expression on their faces.

I was miserable with the cold. I was not unhappy nor lonely; but if you think I was comfortable all the time, you are mistaken. My hands were so numb I dreaded taking off my mitts for that shooting contest. I hated picking up that icy rifle. My cracked, bleeding face was quite as painful frozen as it would have been unfrozen. What I am trying to say is that freezing one's face afresh does not necessarily deaden all the nerves. And the wind had a beastly way of shooting up under my *kooletah*, making my back and shoulders so cold I sometimes wondered if I might not be in danger of freezing to death. The thermometer stood at forty degrees below zero.

In mid-afternoon we came upon our first caribou tracks. They were a week old, and had been more or less drifted over. We decided there must have been about fifteen animals in the band. They had been travelling slowly northeastward.

We built a big *igloo* that evening. It was about twelve feet in diameter inside, and was made of more than sixty snowblocks. To the east of us lay East Bay, a black ribbon of open water in the middle smoking fiercely. Beyond the bay rose the bold headland of Gore Point, and, farther on, above the rugged hills, the dome of Mount Minto. Wild country this: home of *Tooktoo*, the Barren Grounds caribou; home of *Amaughuk*,[1] the Arctic wolf.

At supper time we talked of nothing but caribou. It was *tooktoo* this and *tooktoo* that; antlers, hooves; scent glands; reindeer moss; deerskin sleeping-bags; young, old, summer, and winter *tooktoo*; and *tooktoo* migrations. Never was there in the history of Arctic exploration a more concentrated symposium on caribou. The little island just north of East Bay was, I learned, called *Tooktootok*.[2]

After supper we "conversed" upon sundry matters other than caribou. Muckik tried some English words. He failed so badly in his pronunciation that John laughed loudly at him. I tried some new Eskimo expressions and we all laughed. The dogs evidently believed something wrong, for they started to howl. But they stopped when John shouted *"Palaghit!"* at them. *Palaghit* is a dignified imperative form, but when John used it on the dogs it took on the purple hue of all effective cuss-words.

During the following several days we hunted for *tooktoo* almost constantly. Trailing caribou on lightly loaded *komatik* was lively business as compared with our laborious journeying from the post. Now we rattled and bumped along, keeping ourselves as comfortable as possible on the thin bearskins that covered the tops of the sledges, spending about half our time, I should say, in mid-air between bumpings. I was glad to leap off for a run now and then, but I never let the *komatik* get far ahead of me.

We made one trip far into the interior, reaching a high hill whereon we built a pyramid of stones. From this lookout we surveyed our surroundings with binoculars. We did not

---

[1] Barren Grounds Wolf, *Canis tundrarum.*
[2] Place where caribou live.

sight any caribou. But eventually we came upon many signs of the animals: broad tracks in the snow; diggings and scrapings among the moss; and innumerable round droppings that struck me as surprisingly small for so large an animal. We made our way out to Tooktootok Island, where we spent two days. I walked the entire length of Tooktootok, saw many caribou signs, and decided that the herd, numbering perhaps a hundred individuals, must have crossed the ice back to Shugliak, and now be somewhere to the northwest of us among the high Fox Channel hills.

Since our supply of dog-food was low, we shot some *netchek*-seals. At the "holes" the odor of musk was so strong that I could smell it easily when I knelt and put my nose close.

One morning we were making our way across a large lake. The dogs had been directed to go straight ahead, but they kept veering off to the right. The ice was so smooth we were comfortably seated, for a change.

John turned to me and said, pointing toward himself as he spoke: "Tell dogs fin' *tooktoo*, she unnastan!"

John's one and only English pronoun was *she*. He could not remember the word *I*; and he pointed to himself by way of letting me know that when he told the dogs to find the caribou, they understood him. I sensed vaguely what he meant in an instant. But full understanding came within the next half hour.

John hissed two short words at the dogs: "*Huit! Huit!*" Just two short words, not spoken loudly. Two small explosive sounds and nothing more. But the dogs went wild. The slim bitch-leader turned her head for one reassuring glance, then made off at high speed, followed by her eager comrades. The *komatik* leaped, jerked and clattered. The runners screeched. We swung back and forth dizzily on the snowless ice, and scattered the thin drifts in sheets. The team paid little heed to John's minor commands. They had been told to "fin' *tooktoo*," and they would do their best. Suddenly the bitch-leader paused. The team slackened their pace. We kept the *komatik* from running into them by digging our feet into the snow. There was a jerk as the dogs all changed their course sharply and with an altogether unexpected access of energy doubled their speed. The *komatik* all but overturned. Before I knew it I had been thrown off, for all the world as if we'd been playing a wild game of Crack the Whip with me at the nether-end.

When I caught up with the expedition once more, John assured me that the caribou could not be very far away. He had signalled to Muckik and the other *komatik* was coming toward us. Within a short time the dogs had been anchored securely to boulders, the *komatik* overturned, and the bearskins rolled in a safe place out of reach.

We started afoot toward the clifflike edge of the island, continuing in precisely the direction the dogs had brought us. Walking was easy on the lakes, but climbing the jagged rockwalls and crossing the glassy drifts were hazardous. I couldn't help being skeptical about these caribou. We had covered so much country by this time, and had so many times been disappointed, that I was in a not very enthusiastic frame of mind.

Finally we found ourselves unable to go farther. We were at the brink of a two-hundred-foot cliff. Below us stretched the rough shore-ice and the smooth deep-water ice of East Bay, and beyond this the whole forlorn length of Tooktootok. We turned southward, making no attempt to descend the cliff.

As we were skirting the crest of a little knob John suddenly crouched. Muckik and I ducked in an instant. On a sheltered slope a quarter of a mile inland were ten caribou resting in the snow, most of them apparently cows, one a fine bull with towering antlers that made me think of a lightning-struck tree.

We crawled quietly forward a long way, not daring to peer again over the knob's crest. But we had been careless in our first approach. When next we sighted the band they were on the run, much farther away, headed for a narrow plateau to the westward. They stopped for a time in the middle of a lake. Here the big bull pranced about nervously, sometimes standing high on his hind legs the better to look at his domain. His actions made me think of a gigantic hare.

Eskimos do not like to walk, you remember. It is a matter of solemn fact, however, that we three *tooktoo*-hunters spent the following three or four hours walking and running and climbing about on hands and feet in direct, personal pursuit of those ten caribou. We gave up all thought of a noontide mug-up. I was thankful for my long legs and glad I had been walking enough to be in fair form. I tried once or twice to ask my companions just what we were planning to accomplish as a result of all these simple, cross-country maneuverings, but John could only make motions with his arms and do his best to tell me that we wanted to get as many *tooktoo* as possible. With this idea I concurred, so we proceeded to chase the beasts.

There, only a little more than good rifle-range ahead of us, and frequently in plain sight, were those big, strong-legged caribou, moving forward in a series of swift spurts of speed and nervous pausings. We followed as fast as we could, running on the down-slopes and level stretches, toiling up the hillsides and ridges. We must have run and walked and climbed all of six miles. Night was descending. I was wet as a rag and found myself wondering if this overheated condition might not be dangerous in the bitter coldness. John and Muckik also were hot and wet; and they panted noisily. But we kept on running. I was tired, but the longer I ran the more interested I became in this odd enterprise. This chasing down a band of caribou was novel sport indeed. How soon might we hope to catch the animals? Would we tie them up with ropes? Would they kick us or bunt us when we tried to hold them, or should we have to hamstring them with knives? I couldn't help thinking of the days when, as a child, I had stumbled across newly ploughed fields trying to put salt on a blackbird's tail.

The sun had set. We stopped for a rest. John now informed us in signs that he would go to one side of the ridge toward which he pointed; that I would go to the other side; and that Muckik would go straight over. I sensed some sort of definite plan at last. Now that night was at hand, the animals would be seeking shelter. We started again.

Soon Muckik was signalling me from the top of the ridge. I could see his arms waving in the dim light. He came toward me, saying no word, but pointing in the direction John had gone, then in the direction of the teams we had left so far behind. Muckik knew one or two English words, but he could not make me understand his English whisperings. We tried whispering in Eskimo. He whispered as slowly and as distinctly as he could, and I did the best I could at listening. We stood there looking steadfastly in the half-light at each other, trying to make sense out of the half-sounds. It was all but hopeless.

Finally I thought I understood. John had gone on, for he had sighted the *tooktoo* and apparently knew where they were planning to huddle for the night. And Muckik and I were to return to the dogs, drive the teams back to camp as soon as possible, and wait there for John.

For a moment I was angry. It would have given me a certain satisfaction to shoot that big bull caribou myself. It would have given me great pleasure to observe the animals, or to watch the Eskimos stalk them. But how could I express these thoughts to my friend Muckik? Or how could I be sure that I had understood his earnest motionings and frantic whisperings correctly? Furthermore, I was tired. After all, this was Eskimo country. Muckik and John were the real *tooktoo*-hunters; I was only an interloper. This *tooktoo*-hunting was part of life on the tundra. For me the tundra, all the *tooktoo* on the tundra, everything about the tundra, everything about the whole Innuit world, in fact, was only an interlude.

But this matter of driving the dogs back to camp. *Me* drive John's dogs? *Me?*

Muckik and I started back to the teams. I was silent. I didn't even try to say anything. The darkness all at once seemed to bode some evil. This was no time for asking further advice in dog-driving. What I had learned I had learned. If the gods were with me, well and good.

I think I hypnotized myself to some extent as I walked back across that stretch of tundra. *You must not be afraid of those dogs,* I was thinking. First of all you must not be afraid. If you are afraid the dogs will sense it immediately. You dare not count on simply acting as if you are not afraid. The dogs will know it if you try any make-believe on them. You really must not be afraid. All the way back that is what I kept thinking to myself. None of this keep-your-chin-up stuff; none of this stiff-upper-lip stuff; something more basic than that.

By the time we got back to the *komatik* I may or may not have been afraid of John's dogs. I believe I wasn't afraid of them. I doubt if I had ever been really afraid of them. For one thing, I had never been attacked by a Husky, so there was no disagreeable recollection to conquer. Too, the dogs all knew me to some extent.

Muckik gave me no word of direction. He simply turned over his own *komatik,* unwound his dogs' traces from the boulder hitching-post, and started off.

As for most of what I did the less said the better. I am not proud of my abilities as a dog-teamster. Nor do I wish you to think that dog-driving by beginners is always as successful as this attempt. I marched in amongst the team as one having authority. I knew at least three of the dogs rather well, and the peculiar inflections used in addressing these dogs I remembered clearly. I turned the *komatik* over, lashed on the bear-skin, and disengaged the traces.

Kopernoak, a tawny brute, eyed me with suspicion. "Aha!" he quite obviously was thinking, as he looked at me with his small, colorless eyes. "Here's a promising situation indeed. Anyone can see you're not used to driving a dog-team, but you're acting very almighty about everything as if you expect us all to obey immediately." I knew only too well what was in Kopernoak's mind.

I picked up the whip, kicked loose the front of the sledge, and rapped with the whip-handle. The team were up in an instant. They stretched and whined a little, then bounded off, glad to be on the move. But Kopernoak did not pull an ounce of the load. He sauntered along for a quarter of a mile, then turned insolently and yelped at me.

"Kopernoak!" I shouted in my roughest voice. "Get along there!" Instantly the team stopped. Something was wrong. This language, this voice, they did not know. I would gladly have hidden my face.

The traces were so badly entangled that one lame dog had difficulty in keeping his position. I decided to take a breathing spell while rearranging the team. When I leaned over to straighten out the traces, three dogs, led by Kopernoak, bounced back to nuzzle me. I could feel their gentle mouthings, but I paid little attention, merely wiping them off when they got in my way. I realized then that I might become afraid of them, desperately afraid in fact, were they to seize me by the leg and start to drag me down; I wondered vaguely how quickly I could get my rifle; but I went on straightening out the traces. After twenty minutes of fumbling with cold hands in the dim light we were ready to start again.

Again I rapped with the whip-handle and the team rushed off. This time Kopernoak deliberately lagged, his trace so loose that it caught under the runners. I could see that he was daring me to use the whip. I evaded the issue by flicking the loose trace about his head. This had never before happened to him so he set up a muffled barking. A Husky, you recall, almost never barks. At once the team stopped. The bitch-leader was an obedient worker, but her personality was not always as powerful as Kopernoak's. I saw that it was time to take drastic action of some sort. I rose from the sledge, snatched up the whip and, with the heavy handle, whacked Kopernoak on the nose. He whined; gave a bewildered yowl; and fawned at my feet. I probably had not hit him harder than any Eskimo boy would have hit him merely as a passing gesture; but I had hit him harder than he thought I would, and he behaved differently from that moment on.

The lame dog continued to limp and to get his trace badly tangled. Finally I stopped the team again, took the lame dog out, and tied him onto the *komatik* alongside me, under part of the bear-skin. He was little more than a puppy.

Our journey to camp across the aurora-lit snow was not without other difficulties, and the runners, I must confess, grated viciously over many a sharp-edged stone. But we finally made the *igloo*, and as I freed the dogs from their traces I sensed a new friendliness in their nibbling at my mitts and boots, their impetuous half-growling, and their eagerness to declare internecine warfare.

Muckik lighted the Primus and Coleman inside the *igloo*. Soon we were comfortable, warm, and drowsy. It had been a busy day and we were glad for a rest. Neither Muckik nor I had caught cold as a result of being overheated. We tried talking to each other, but my Innuit was nothing short of lame, halt, and blind, so we sat there wondering about John and—to use John's quaint phraseology—"thinking about animal," while tea boiled. We had a mug-up. This gave us a chance to express our satisfaction in the sounds and facial expressions of a language we both knew. Then we lay back and mused for another hour.

All at once there was a low rustling, and a whining among the dogs. Muckik sat up, head cocked, mouth open. There was a faint crunching of snow, then a loud thump on the *igloo's* roof. One thump, two thumps, three thumps, four thumps! And Muckik let out a fierce cry, a cry so wild and so loud that my hair stood on end.

I may as well confess that those four eerie thumps and Muckik's piercing yell had me genuinely terrified for an instant. What manner of visitation was this? If some human being were outside, why had he not spoken? If it were some beast, why had the dogs not made attack? Were we, at last, in the dread clutches of the evil spirits? Was Muckik going *problokto* [1] before my very eyes? Had we broken some ancient and sacred taboo? Were the gods of the *tooktoo*-hunt offended?

There was a grating sound along the floor as the big snow-

[1] Mad.

block at the entrance moved slowly back and forth and finally
inward. A furry, shapeless mitt showed at one side. A smiling
face peered in. It was John.

John was back from the hunt. He had killed four caribou.
He had brought the four hides back with him. Four carcasses
were lying in the snow not more than a mile from camp,
ready to be hauled in on the sledges. We had deliberately
chased those animals to a certain valley where the Eskimos
knew they might be shot at nightfall. I had helped in the
chase without more than vaguely knowing what we were
doing. We would now have all the *tooktoo-quak* we could
eat. And there would be hide enough for a fine new sleeping-
bag and a big *kooletah*.

But how was I to know that solemn thumps on an *igloo*,
unaccompanied by so much as a single spoken word, is sign-
language of the successful *tooktoo*-hunter old as the race of
Aiviliks?

CHAPTER XXVI

A Day of Thanksgiving

So AT last we had four caribou. The hunt had been a long one, whole days of bumping over lakes and drifts, winding up with a never-to-be-forgotten cross-country Marathon. I was so weary that night and my hands so swollen and stiff that I wrote nothing in my diary. But next morning before breakfast I jotted down some lines and suddenly realized it was Thursday, November 28: Thanksgiving Day back home in West Virginia!

By way of celebration I made two sketches: one of a gobbler strutting, the other of a roasted turkey on a platter. These I showed to my friends, explaining that they were both the same bird, one before, so to speak; the other after.

Both men laughed. Muckik asked why the strutting gobbler had no feathers on its head.

This question, as you will perceive, I did not answer very cleverly. The idea that gorgeous bareness of head might to the turkey mind embody all that is truly beautiful and worth while in the world was an idea a shade too esoteric for our *igloo*. I evaded the issue by answering that turkeys live in such a hot country they probably do not need any covering for their heads.

At this the incorrigible John chuckled, pointed to the picture of the turkey on the platter, and said: "She plenny *ikki*, sho?" [1]

---

[1] "She plenty cool, so?"

The wind was strong, the snow drifting; a blizzard was blowing up. We piled out into the raw air, harnessed the dogs, and raced off after John's caribou. The blast was so fierce I had to hold my hands over my nose carefully to keep it from additional freezing. Finally, resigning myself to the discomfort of breathing through thick, ice-impregnated wool, I pulled my Nansen-cap down over my face.[1]

We had trouble finding the carcasses, for drifts had almost buried them. John had killed the big bull, two cows, and a youngish male animal. He had skinned them immediately, tearing out the entrails and laying these in heaps to one side. Here foxes had been gnawing and scratching. We set some traps on the intestinal masses, but piled the stomachs on the sledge along with the meat. We took care not to damage the antlers of the big bull, for I wished to preserve these and the skin as museum specimens.

On the way back to camp we noted bare spots where we thought caribou had recently been pawing; but we saw no caribou. Here and there we traversed wind-scourged areas where old trails were oddly perched on thin ridges and slender stalks of snow. The wind, unable to wear away the firmly packed snow of these tracks, had carried off the surrounding soft snow, leaving pretty colonnades where foxes and dogs had run, and airy causeways where our *komatik* had passed.

I was glad to get back to the *igloo*, for I did not like the idea of being lost in this storm. John said we were to have one of those five-day blizzards.

I was genuinely fond of frozen raw caribou meat by this time, and downed quantities of it at every meal. Of *toonuk*,

---

[1] The Nansen-cap (named for the explorer, Fridtjof Nansen) you may wear on top the head in the manner of an ordinary cap. Or you may wear it with the sides pulled down over your face and ears. There is an ample opening above the nose through which you can see; but you have to breathe through the cloth in front of your nose and mouth. Moisture gathers quickly in the wool. This freezes and you find yourself wearing a sort of ice-helmet.

the fatty part, I was not so fond, for the stuff adhered to the roof of my mouth and the lining of my gullet. Nor had I learned to like the cylinders of frozen marrow that we cracked out of the leg-bones. This marrow the Eskimos liked best of all.

I experienced an odd sensation when I saw Muckik gouging a frozen eye from its socket in one of the skulls, nibbling the fat from round the ball as daintily as a mouse nibbling cheese. I was almost bewildered when I saw my companions chopping out portions of "salad" from the frozen paunches. This partly digested mass of moss and grass and bits of twig and bud they mixed with *netchek*-oil and ate enthusiastically. Determined to learn what I could about this somewhat mysterious matter of being an Eskimo, I tried a little "salad" too, but decided that certain mysteries were more attractive unsolved.

I spent the day measuring carcasses. Since I had left my tape at the post I had to record these measurements by tying knots in strings. John had skinned the caribou very crudely, cutting across the faces, through the mouths and eyes. But I foresaw that I should be able to preserve the complete specimens eventually if only we could keep the dogs from chewing at the heads and feet.

There was a long and unexpected discussion about dreams. John told me he hunted almost every night for bears and wolves and whales. Muckik told us (that is, he told John and John told me) that he, too, hunted every night. Apparently the Eskimos regarded their dreaming not as a psychological phenomenon, but rather as actual happenings in the life of that part of themselves which awoke, and walked forth, and hunted by night. Jack Ford, I remembered, was always telling me his dreams back at the post. At breakfast we had played interpreter morning after morning. But here the dreams were matter-of-fact stories about a man who lived inside you, a man who slept while you were awake and who awoke when you went to sleep.

We had yet another symposium on caribou. The antlers of our big bull were still partly "in the velvet," little strips of plush-covered skin clinging to the long tines and to the corrugated bases. John said the caribou scraped this "velvet" off on stones. He told me he had seen the *tooktoo*-bulls fighting fiercely in October and early November, and that he had found many shed-antlers of both cows and bulls. But when I asked if he had ever found any locked-antlers [1] he appeared to think my question purposely humorous.

KILLED BY THE WOLVES: THE SKULL OF *TOOKTOO*, THE BARREN GROUNDS CARIBOU

I remembered what Sam had told me about the hordes of "deer" that had swarmed about the inlet back in 1924, the year the post had been established. "Deer" had been everywhere that year, unsuspicious, docile, friendly almost as cattle. The Eskimos had killed great heaps of them. But now

[1] Antlered animals sometimes shove their heads together in such a manner as to cause the antlers to lock. Unless such animals pull themselves free they eventually die of starvation.

the *tooktoo* were to be found only in the wilder areas, along the plateaus between Duke of York Bay and the cliff-lined shores of Fox Channel.

Conversation suddenly centred upon the word "pots-poe." John said: "White Man, she shay pots-poe." Then, pointing to himself, he continued: "Husky,[1] she not unnastan *pots-poe*."

I perceived that John wished to know what the White Man meant when he used the word *pots-poe*. But what was this word *pots-poe?* This was one of those times I must do my best. An Eskimo whom I liked very much was asking me a direct question about my language. I must not fail.

This word *pots-poe*. Or was it a word? Perhaps it was a phrase. Perhaps it was a whole sentence. Was it a noun or a verb? I thought of *passport, passe partout, participle, participate, potatoes, postpone,* and dozens of others. But these scarcely were the sort of words John would be asking about, and none of them was the one he meant. We must have hashed the matter over for half an hour.

John gave me a lead, finally, when he said: "White Man, she shay something *pots-poe*, something not *pots-poe*."

I had it at last. The word was *possible*.

This problem solved, I chanced to remember that I had brought with me a can of specially selected "French" peas. What better day could there be than Thanksgiving Day for eating these?

I dug the can out of my belongings; but we could not find the can-opener. So John set the can on the Primus, saying that he would open it with a hatchet after the contents had thawed a little.

We returned to caribou-measuring and conversation and

[1] The word *Husky*, when applied to a person or to a race of human beings, is slang. It is not slang when applied to the Eskimo's dog. And it is perhaps not quite slang when applied loosely to the Eskimo language.

forgot all about the peas. No odor of burning assailed our nostrils. No sound of boiling gave us the slightest warning.

All at once there was a furious *powf!* as the top of the can blew off; a fierce hissing as scalding peas rocketed to the ceiling. Fortunately no one was injured. The dogs began to howl.

A very few peas remained in the can. These we ate with ceremony, thankful we had not been shot in the face with them.

And during the rest of the evening we hunted peas, finding and munching them one by one. Peas that had but so recently been on fire. Flaming peas that had shot heavenward. Peas that had dreamed, perhaps, of becoming stars, but that had wakened to find themselves gone suddenly cold: mere planets in a low and miserable mock-heaven. Mere peas, after all, scattered all over a snow-house roof.

Streamlined tupek, or skin tent of the Aivilik Eskimo.

Skull of one of the extinct Shugliamiut.

A Peregrine Falcon on nest.

A male Rock Ptarmigan.

Aivilik Eskimos and sledges.

An Eskimo icing the runners of his sledge.

Eskimo companions Muckik, with the doctor's binocular case, and John Ell.

Muckik pointing out caribou country to John Ell.

A sledge loaded with caribou.

Arctic fox caught in a trap.

"Viscount Grey" a pet Collared Lemming.

Weasel.

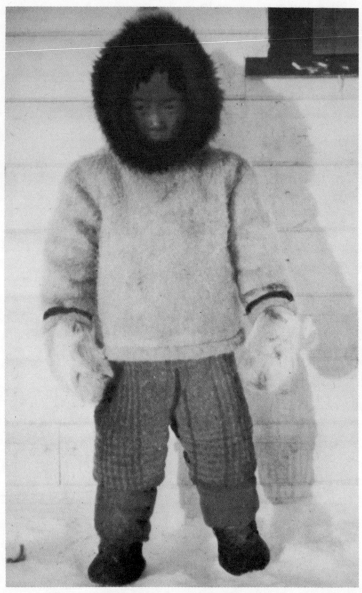

Little Peter, whose Aivilik given name is unmentionably obscene.

John Ell and Muckik.

Aivilik girls at the entrance of their igloo.

Chapter XXVII

Storm-Bound

THE sun probably rose over some far corner of the world on the morning of November 29, 1929. But I have my doubts.

By nine o'clock I was thinking of getting up. The wind was sullen. All night long it had filed away at our *igloo*, snarling and growling at any object that resisted its mad careering. A drift completely filled the space between our hut and the low ridge to the north, and now, to either side of us, an immaculate wing of snow swept down and outward to merge with the horizonless expanse. It was dangerous, on such a day, to hunt the caribou. It was impossible to see even a few yards through that giddy "whirl-dance." To face that wind, to expose ourselves to those driving tacks of snow, would be vain, bitter torture.

I thought of the story Sam Ford had told me of an Eskimo who had left his *tupek* in such a blizzard, heading for another part of his village, a few rods away. The unfortunate man had missed the path somehow, and had wandered farther and farther through the storm. He had been lost for nine days. Finally the dogs had found his body, frozen stiff, buried in a drift.

At ten o'clock I thought again about getting up. These ancient traditions of the diurnal: they would somehow have to be upheld. I found myself regretting that human beings do not curl up and sleep away the winter along with mother-bears and chubby little *Shik Shik*,[1] the ground squirrel.

At eleven o'clock I roused myself, tousled my hair furiously, and pounded and rubbed my arms by way of making commotion and getting the day started.

John and Muckik stirred, commented upon the wind and condition of the *igloo*, and reached for their pipes. After indulging in a long smoke, and by almost imperceptible stages, they emerged from their wrappings and dressed.

Muckik's face broadened with a smile as he shouted *"Piuyuk!"* [2] He had thought of breakfast at last. He inspected the Outer World through a tiny hole he poked through the *igloo's* wall, lifting his left leg comically as he strained and squinted his eyes. Obviously it was to be a quiet day at home.

Breakfast over, coffee grounds flung to the ceiling, spoons restored to their box, we sat about peacefully, no one grumbling that the day was not otherwise.

Muckik observed that the supply of fat in his *koodilik* was low. He uncovered a frozen *netchek* and, with his *pana*-knife, cut from its belly a big rectangle of skin to which adhered an inch-thick layer of fat. Turning the slab inner surface up, he pounded the stiff blubber to a pulp with a stone. This mass he spooned into the shallow basin of the lamp where it spread out and, encountering the moss-wick at the edge, filtered upward into the serried flame.

The heat from our several fires so increased the temperature that the roof melted rapidly, great drops of chill water plopping on our heads and necks and amongst the bedding. *"Igloo kootook-took!"* [3] shouted John, as he tore pieces of

---

[1] Hudson Bay Spermophile, *Spermophilus parryi*. This animal is not found on Shugliak; but it is common at Repulse Bay and the Aiviliks therefore are familiar with it.

[2] "Good!"

[3] "The igloo is melting!"

paper from tea packages and pressed them gently against the half-formed drops.

The unpleasant rain continued, however, to descend. We dug troughs in the wall, stuck chunks of snow up to act as sponges, and turned off the Primus. Our faces were wet with perspiration. Great holes gaped in the roof and snow sifted in with the icy drafts. John rose, put on his outdoor rigging, took his *pana*, and slicing the door from its "hinges," made his way outside. The candle flames wavered and tossed, but burned more brightly as fresh air rushed in. I drew on heavy clothing in haste while Muckik cleared back the bedding in preparation for the period of repairs. Soon new blocks, dazzlingly white, appeared above us, and though heaps of snow and wet ice fell in as the work progressed, we were soon snug once more.

I wrote a few notes in my diary-book. But both my mental and my physical processes were sluggish and I preferred to sit, hands interlocked behind my head, musing on this incredible situation: here I was, storm-bound in an *igloo*, two days away from the post, sleeping alongside the Eskimos, sharing their raw caribou-*quak*. Home, college days, friends who spoke my language: these seemed far, ever so far, away.

I was roused from my reverie by Muckik who, having surveyed me out of the corner of his narrow, close-set eyes, inquired shyly if I'd like to play some *"Fï' Hunnun."* [1]

When I assented he produced a brass-edged box, and having opened it, began a ten-minute search through its chaotic depths. One by one the playing cards came to light until the entire deck was assembled. Never, of course, were there cards so black and foul. In the soft light I could scarcely make out the figures on their faces. Ragged-edged, flop-eared, cracked and oily, they stuck together so closely they were scarcely to

---

[1] Five Hundred.

be dealt. But by degrees I found it possible to recognize certain cards by other features than color, numeral, or design. The Ace of Spades was not one whit blacker, in this Aivilik world, than the Six of Diamonds or any other card, but I learned to recognize it when it came my way, by a scratch across its face.

My friends had learned at the post the rules of three-handed Five Hundred. Theirs was, however, a glorified game, with two Jokers, doubling, redoubling and indefinite bidding. They asked me to keep a score, though I soon saw that this was suggested only to give me added amusement, for the score had nothing, absolutely nothing, to do with the game.

The delight these child-men derived from playing without a thought of winnings or losses made me envious. At first I simply could not catch the spirit of their game. I was bent on winning the coveted five hundred points. By the time Muckik was "in the hole" some three thousand, seven hundred points, however, I saw that were I really to *play* with my friends I should have to forget the score, though I marked it down.

I learned the many and subtle rules. I found that I, too, could leer across the table and hiss a bid between my teeth. I, too, could dramatically hurl my hand at the table and shout *"Piyungituk!"* [1] By degrees I perfected the *sotto voce* that portended higher bidding later on. Finally I could slip my cards in slyly, gloat over a trick won, or fling down a useless small spot quite as artfully as my companions.

We played five hours at that sitting. Muckik's "losses" ran so far into the thousands that I hesitated to waste our precious pencil in continuing the record.

"Shick calub!" he roared, displaying a lonely upper tooth

[1] "Bad!"

in the downward curl of his lips. Bids were made altogether in understandable though distorted English.

"Shick dima!" countered John, his eyes narrowed, his breath coming heavily.

Snorting contemptuously, I passed.

Muckik surveyed his hand. "Shick No-Tlunk!" he crowed, pounding the table.

John was not to be downed. "Savong Shipi-ed!" he shouted.

I glared at my opponents. They looked harpoons at me. They knew I was about to make a desperate bid. "Nullo!" I fairly shouted. The faces of my adversaries dropped for a further survey of their cards. "Nullo," in which the bidder may not take a single trick, is considered a very smart bid. After several moments of profound thought, the Kitty, four unknown cards that go to the highest bidder, was conceded to be mine.

All at once John and Muckik had a physical squabble over a card that had been misdealt. John jerked and Muckik jerked, and the card was torn in two. What a calamity! It was the precious Ten of Diamonds, an important card in a game they called "Kahcheeno." [1] Instantly our playing stopped. With a solemn face Muckik rooted out his brass-edged box, and dug from its confusion of tools, nails, trinkets, buttons and pieces of ivory, a heavy needle and a dry strand of *eevaloo*. [2] As he forlornly picked up the pieces of his damaged possession a wave of sympathy flooded me; what would I not have given, just then, for a brand-new deck of cards!

With painful precision he sewed the card together. It was ruined, in a sense, for it would now be recognizable from

---

[1] Casino.
[2] Caribou-sinew thread.

the back. But Muckik would do his best. The stitches were neat and small. He passed the needle straight through the paper so as to make his mending as strong as possible, but now and then a stitch tore through and his brown face clouded with dismay.

Our spirits were somewhat dampened by the misfortune. Muckik put the cards inside, returned needle and *eevaloo* to the brass-edged box, and, as if eager to think about something else, turned to ask me my name.

John answered Muckik's question for me, saying that my name was *Doc* or *Doctor*. I had now to explain that neither *Doc* nor *Doctor* was my real name, but that my real name was *George*.[1] This momentarily confused them. I tried to explain that the name was a combination of two words from an old, old language, and that it meant "Tiller of the Earth," or "Digger of the Ground." I suppose I shouldn't have been so pedantic, but conversation of this sort was interesting. It might lead anywhere.

My real name, then, was "He Digs in the Ground." The light of understanding slowly dawned. Here, at last, was partial explanation of my strange behavior! No wonder I had been unearthing specimens of plants; no wonder I had been excavating burrows. If my name was "He Digs in the Ground" why should I not be a gatherer of grass, or a catcher of lemmings?

Perhaps over-eager to impress my friends, I explained that George was only one of my three names. Again their faces showed interest—and bewilderment. John was still in some doubt even about this strange name *George*, this strange "He Digs in the Ground," until I told him that the name was common in the South Country, and that it must be a perfectly good name since it was the same name as that of his

[1] Sam and Jack always called me "Doc," or "the Doctor," never "George."

very great King, who lived across the *imakjuak*[1] in England. My friends now told me of their names. John's real name was Amaulik Audlanat, meaning something like "Dorsal Fin of a Salt Water Trout He Goes Away." At the post, Amaulik had years before been nicknamed "John L.," after the famous pugilist, John L. Sullivan. So apt was the sportive title that Amaulik was now widely known as John Ell; indeed he was sometimes referred to as Mr. Ell; I had even read about "Chief" *John Elliot* in a Montreal newspaper. After all, my names were in no worse a state of confusion than John's.

Muckik's name meant "Something Rising Up," like an *igloo* being built, or *Nanook* standing on his hind legs. He had another name, too, but I couldn't write it down, nor pronounce it, nor learn what it meant. It was pronounced somewhere between the jaws and the stomach. Its jumbled consonants were too much for me.

Muckik, whose active brain seemed ever eager to corner and master a new English word, now began to practise on *bird*. I could not forgo the relief of a smile, even a quiet laugh, at his efforts. Pouting hideously, his great lips forming the syllables, and his tongue, palate and overworked teeth doing their utmost, he finally delivered a suggestive "bris-dl," the *r* rolled considerably, the *s* produced against the rear part of his gums. His next trial resulted in a frank, loud "broad-rr." Once more, and it was "broad-l"; again, and he floundered on a "bladz-rrl" which seemed almost to down him. But he kept on until he finally got it "blarrd," which, considering his dental deficiencies, and his utter unfamiliarity with our *r*, was not so bad.

Having spent a profitable hour in this side-study of linguistics we ate a late noon-day meal of *quak*, hard biscuits

---

[1] Big water; ocean.

and tea. Stretching ourselves, we peered out at the storm, shrugged our shoulders, and returned to our stations among the caribou-skins, thankful that we were inside. Night had descended, but our candles burned, and we were happy. The cards came out again—even that war-scarred veteran, the Ten of Diamonds. There was another three-hour session of dealing, bidding and playing. The scores, which now completely covered the inside of a pasteboard sugar-box lid, ran into four and five digits. I could not neglect this, my special part of the game, without being called to order.

Suddenly, as he was shuffling the cards, Muckik began to sing. The tune was vaguely familiar, and I noted that John joined in. By the time the air had been repeated three times I recognized it as *Nearer, My God, to Thee*. At first I was afraid to join, fearing that my voice would embarrass the Eskimos. Presently I began to hum, ever so lightly, stealing a small glance at Muckik, who had been looking at me out of that twinkly corner of his eye. The song continued, this time with words learned at the Mission.

Muckik continued to shuffle the cards, but his soul was plainly in the music. We all looked at the table, but sang now, without restraint. Remembering that I had once been able to harmonize without disaster in simple tunes, I wandered into the baritone and tenor parts. The lowly dome of our crystal cathedral rang with sturdy chords.

By the twelfth repetition of the melody Muckik had slowed his shuffling of the cards down to a slow movement almost metronomic in its regularity. His hands moved automatically. But his rich loud voice carried with it the enthusiasm of his innermost being. John, too, was singing loudly and well, and I, at last unafraid to express myself, was employing all the ability at my command in rounding out the theme with new variations.

Little by little Muckik's hands ceased their mixing of the dirty, worn-out cards. After the twentieth repetition of our hymn we stopped. We had sung the four stanzas through five times. Our game of "Fi' Hunnun" was at an end.

Just before I went to sleep, I noticed that Muckik was squinting at something he held near the yellow flame of his candle. He had somewhere found a small, transparent piece of paper, and was tracing from the sugar-box lid the figures I had made in keeping that interminable score. What was his purpose? Had the score after all some connection with our game? Or was there inherent in it some magic to which I was blind?

CHAPTER XXVIII

Akjuk: the Season of Shortest Days

JOHN, Muckik and I spent four days in our storm-worn *igloo*. We crawled out to view the weather now and then, mended the roof several times, and fed the dogs twice; otherwise we were inside all the time. We shredded a bundle of sinews from the backs and necks of the caribou into stout *eevaloo*-thread. We talked a great deal, sang songs, had some wrestling matches and finger- and thumb-pulling contests, mended clothes, cleaned rifles, wrote in our several diary-books, and consumed prodigious quantities of *tooktoo-quak.*

We had an involved discussion as to what the White Man calls *A Year.* It all started when I asked John how old he was. John, being a good Eskimo, answered in what he considered the most companionable way, raising his hands dubiously three times because he thought me about thirty. He had followed the same tactics with Sam at the post, placing his age that time at fifty, so as to be companionable with Sam.

We found ourselves all tied up in Arithmetic when it came to the three hundred and sixty-five days in the year. There was all manner of counting on fingers, moving of piles of buttons, and tying of knots in strings. The significance of all this counting and moving and tying I did not catch; but apparently it helped the Eskimos to clarify their concept of the thing called *A Year.* I had no end of trouble in explaining that we were at the end of our month called *November* and at the beginning of our month called *December*. Muckik had a little calendar I had given him. On this he crossed off the days as they passed, with a pencil. I rattled off the ditty "Thirty days hath September" for the men, and there was a hearty laugh. Muckik informed me that he wanted to write the ditty down in his diary-book sometime *"wacherapik."* [1]

John told me the Eskimos called the season of the shortest days *Akjuk*. *Akjuk* was also the name of a bright star that could be seen at this season just before dawn.

You must remember that winter on Shugliak is not a sunless season. If you are thinking of winter in such terms as "six dreary months of darkness" you are badly mistaken. The sun rises and sets all winter long on the Arctic Circle. You see the sun *every day* unless there is a blizzard. But during December the sun is visible for such a short time each day and it is so low in the heavens that you can't help thinking of sunset and twilight and night even at noon. There are long blue shadows everywhere. The snow is pink or purple or orange, reflecting the colors of the sky.

On the third day of December we burrowed out to find our *igloo* practically buried. The blizzard had spent itself. The sky was pale, thin turquoise. We dug harnesses and *komatik* up and fared forth to visit the fox-traps and to look for caribou trails. We found no fresh *tooktoo*-signs, but were successful in getting some *netchek* for dog-food at the southern end of Tooktootok Island.

We saw a wolf running across a lake far in the distance. He stopped a moment to listen, then turned and loped toward the highlands that were his home. *Amaughuk*, terror of the Barren Grounds! We went up to examine his trail. The footprints were almost twice as big as those of the dogs in our teams.

---

[1] "A little later."

John found several foxes in his traps. Muckik, on the other hand, did not catch any. The traps of the two men had been set side by side and it seemed to me that John's success was purely a matter of luck. Muckik was not in the least jealous, however. He laughed as he dug out and reset his empty traps, saying that John had always been a good trapper of foxes.

I made a sketch-map of the region in my note-book that evening. The Eskimos helped me. Many of the pages in this section of the book were scorched because I dried the frozen ink too close to the candle-flame.

We kept on, day after day, trying to find more caribou, but had no success. Dog-food ran low again. We decided to return to the post for more rations and to try caribou-hunting in another part of the island. On the way back we used the duplex-igloo we had built *en route* to East Bay. We had to patch the roof and shovel wagon-loads of snow out, but it was perfectly adequate for the night's stop.

The last lap of the journey was long and wearisome. It was bitterly cold. There was a steady head-wind. I froze my cheeks and chin afresh. We found ourselves, at nightfall, near Itiujuak, about eighteen miles from the post. But we kept on. With the passing of an hour we could see light from one of the windows of Sam's house, just a tiny point of light, but piercingly clear. This I could not help thinking a long-lost and beloved star, shining so clearly and steadily there, almost on the horizon.

Sam welcomed us gaily. John Ell did not kiss his wife, but he pinched the little finger she held out toward him and made a joyful sound deep in his throat. At this Mary Ell smiled in a confused way, blushed deeply, ducked her head, and lisped a word or two softly in her cheeks.

There were hours and hours of tales to tell. We opened a

bottle of pickles and Sam baked some molasses bread. Everyone was optimistic about the winter. Many foxes had been caught, and the Eskimos were coming in with more every day.

In a corner was a burlap sack full of "specimens for the Doctor." This time it was gravel and sand and chunks of rock from Mount Minto and Seahorse Point. You never saw such a heap of junk. The dogs must not have been too enthusiastic about this side-study of Geology. There were bits of garnet, traces of graphite, and flakes of biotite. It was not a very valuable collection in view of the fact that I had gathered a similar lot while at Seahorse. But I went over the pieces carefully; and the next day Kooshooak, who chanced to be at the post, and who had noted my interest, brought me from somewhere a boulder weighing about sixty pounds. When he dropped it on the floor of my workroom one of the shelves gave way, the gramophone downstairs started playing, and the dogs all over the post jumped up and ran in all directions. You'd have thought a cannon had announced the arrival of some government official.

. . . . .

Since I was not shooting many birds these days, and since there was so little daylight for painting, I requested the Eskimos to bring me the skulls of the foxes they skinned. Skulls began to arrive promptly: boxes of them; strings of them; chewed, chopped, and broken piles of them.

. . . . .

Two ravens that had been caught in fox-traps were sent from Native Point, sewed neatly in burlap. I never ceased to marvel that these birds, whose feet were entirely featherless and whose coloration was so distinctly contrary to the general rule, should succeed so admirably in passing the entire year in this inhospitable clime.

TOOLOOGHAK: THE RAVEN

Sheeloo and Tapatai brought me the trunk of a fair-sized willow tree that evidently had been growing in some well-sheltered spot in the interior of the island. It was only a gnarled trunk, perhaps an inch in diameter and six feet long:[1] a trunk without branches or roots, without leaves, flowers, or seeds. It would have made a grand walking-stick for Sir Harry Lauder.

. . . . . .

Two or three wolves had been caught, by this time, in the Cape Low region. The Eskimos brought the skins in not only for trade but for government bounty.[2] Pumyook had got one of these with a gun-set.[3] He told us an amazing tale

[1] Though this trunk was six feet long the tree probably had not stood more than two or three feet high, for these trunks grow along the ground.
[2] Thirty dollars, if I remember aright.
[3] A loaded rifle aimed at a chunk of meat that is connected by string to the trigger. When the wolf pulls at this meat he also pulls the trigger of the rifle, thus killing himself.

about a young wolf he had caught years before. The wolf had been so tractable that Pumyook had driven it home from the trap in which it was caught, harnessed with his dog-team. I never knew whether to believe this tale or not. But it made me think about these reports I continued to hear of wolves that were Husky dogs gone wild.

Chapter XXIX

A Trap-Line Under the Snow

When, in mid-December, you gaze across snow-buried Shugliak, you are sure to say to yourself: "What a silent world! So desolate, so dead!"

You are silent yourself; so silent that you hear the lisping of loose snow blown across the crust, the crunching of ice in the distant tide, and, far beneath you, the burrowing of lemmings.

Drrrrr, drrrrr, drrrrr! comes the sound, clearly now, directly underfoot. And you cannot, as you listen, keep from cringing a little at thought of the dentists' chairs in which you have sat. A burrowing lemming makes just that sort of sound.

You have read about lemmings, of course: those dramatic creatures that march in countless hordes across the Scandinavian wastes, doomed to "committing suicide" by drowning in the ocean. You may recall that a lemming is a small, stubtailed rodent, and that it lives in Arctic America as well as in the northern part of the Old World. You may have decided, as I have, that lemmings do not arbitrarily destroy themselves by the million any more than human beings do, but that they leap into the sea confident that they will reach some land offshore where there is an adequate food supply. Already I have mentioned the exquisitely soft, pale-gray fur of Richardson's Lemming, that bids fair one day to be as highly prized (and as highly priced) as chinchilla fur. Already I have told you that the abundance of foxes in the North Country depends largely upon the abundance of lemmings. Now I wish to tell you the story of my lemming trap-line under the snow.

All winter I had been wishing to see a lemming in its winter coat. I had heard the little beasts in their burrows innumerable times. I had found their remains in owls', foxes', and gyrfalcons' stomachs. I had caught two specimens during the summer. But I had not yet procured a specimen in winter pelage. I wished to decide for myself how this rodent was solving its problem of race preservation. Did it become white in winter, snow-white like the hare and ptarmigan? Did it have a black-tipped, see-me see-me-not tail, like the weasel's? Or was it, as a species, paying no attention to the possibilities of a protective scheme of coloration and achieving biological success in some other, perhaps less subtle, way?

One morning a lemming-trail led me up a snow-bank to a fresh burrow. No trail led away from the burrow. I knew that a lemming had gone that way not long since: that a lemming in its winter coat must be under me somewhere, in that very snow-bank.

I returned to the post, spread confusion in the placid place by asking for a shovel, put in my trapping bag a half-dozen ordinary, small, spring mouse-traps, rattled round the larder for some oatmeal and potatoes, and returned to the drift.

There I began to dig. Not knowing much about the ways of lemmings, I simply chose a clean, smooth spot at a point where I was somewhat sheltered from the wind, not far from the burrow I had found. The shovel's not too efficient edge hacked away at the icy crust, finally effecting a shallow trench. As I dug down I encountered snow strata of varying degrees of hardness. After the first hour of digging I found myself standing in a thigh-deep hole, a high, crude windbreak of snow-chunks to one side. I was so hot I had laid aside my mitts and taken off my outer *kooletah*, though it was a cold day.

After the second hour's scooping, hacking and tossing I wondered whether I had really seen a lemming-trail or merely been the victim of an optical illusion; guessed that I could work in perfect comfort without any clothes save warm boots; fancied that the Eskimos at the post were discussing the various forms of the White Man's lunacy; and noted that blisters had formed at the base of my thumbs. I wrapped a handkerchief round my right hand and set to work again. I tried to make my task the more enjoyable by thinking about the glorious brightness of the sun, the amazing cleanness of the world about me, and the brisk way the wind had of blowing my shovelfuls of snow into choking crystal-clouds that swept round me.

Another half-hour and I was out of reach of the icy blast. Wet with perspiration but stamping my feet to keep them warm, I rejoiced that I was striking at last into the heart of the drift. The shovel clicked on rock. I examined long-buried lichens, determined there was grass-studded turf between the boulders, noted crevices that had, apparently, never been filled with snow; but found no sign of a lemming burrow.

When I returned to the post I announced that after my morning's labors I now was certain of capturing all the lemmings I needed. In my heart I wondered whether, in all Shugliak's vast expanse there might not be but one lemming —and he an exceedingly wily individual.

On the following day I dug only in shallow drifts. I walked about a good deal before striking in, tapping the crust here, kicking at it there, remembering my blisters. This time fortune favored me. After a few minutes' labor I cut into a neat tunnel that followed the ground, winding this way and that, not among nor under the rocks, but through long grass. After removing the foot-deep snow from an area about twelve feet square I counted along the base of these newly formed walls at least twenty openings to an amazing sub-nivian labyrinth of burrows that extended apparently over the entire meadow. I enlarged these openings, baited the traps—some with oatmeal, some with chunks of frozen potato—and set them in the burrows; sealed the entrances with blocks of snow whittled out with my *pana*-knife, and departed, hands half frozen.

"Any mice yet?" asked John.

"Not just yet," I was forced to reply.

But old Angoti Marik, he who had known much of hardship and disappointment and failure, told his comrades that he thought me the sort of man who will get what he goes after!

Eight days passed. Every morning I visited my odd little trap-line and should have been entirely discouraged had not a trap, now and then, been sprung. There was no evidence that either potatoes or oatmeal had been nibbled; but I was certain the animals were running about, and occasionally across, the traps.

And next day I had a lemming! As I lifted the little, deep-furred creature and its trap from the snow, brushed the frost-crystals from its face and whiskers with my mitt, and noted that it was of a species I had not expected, I had a thrill as definite and momentarily as overpowering as that a big-game hunter must feel at putting his foot on his first elephant. I wanted to shout, to run to the Eskimos and show them I knew what I was about. But I forced myself to keep calm, reset all the traps carefully, walked back to the post, entered the house as usual, and tossed my frozen prize, nonchalantly as I could, on the table. The world's best acting requires neither footlights nor curtain.

Then the argument began. Most of the Eskimos actually never had seen a *brown* lemming in winter. Some told me my specimen was only an ordinary lemming, but its fur had not yet changed to the white of winter. When I suggested that Shugliak might have two distinct kinds of *uvinghuk*, differing in fundamental, anatomical respects, I was politely contradicted. But everyone was visibly relieved and happy that I had caught my mouse. Perhaps they expected that, having procured this one specimen, I would bring in my traps, "lay down de shubble," and come indoors to hibernate until June.

But I went on trapping lemmings. Feeling that the animals were not paying proper attention to the bait I had used, I tried raisins. Immediately I began to catch specimens, more almost, than I could use—big females, full of embryos in various stages of development; youngsters with fragile skin, not long out of the nest; an ancient with a broken tail. And to be sure there were two species: one slender-footed, brown, living almost altogether in the meadow country: the animal known in mammalogical circles as Back's Lemming;[1] the other an amazing, strong-muscled creature with double-tipped digging claws on its front feet, and bristly tail: Richardson's Collared Lemming. What a joyous time I had setting my traps on meadow and ridge, digging in the new drifts, following the pretty trails through the fresh snow, discovering the bulky grass nests, learning that the young—which sometimes number seven—may be born at any time during midwinter, even when the thermometer stands at sixty or seventy degrees below zero!

The Eskimos did not distinguish between the two species,

[1] Named for the explorer, Sir George Back.

calling them both *uvinghuk*. And both species, I found somewhat to my surprise, used the same burrows, running about
and living more or less together; but Richardson's Lemming,
the species that turned gray in winter, the one with the heavy
digging claws, did all the burrowing *through* the drifts;
whereas the weak-footed Back's Lemming, the species that remained brown throughout the year, made tunnels through
the grass *under* the snow rather than through it.

Lemmings! How as a lad I had thrilled at reading of
these animals: their steady march across the tundra; the
frightful devastation wrought by the ravenous hordes; their
plunge into the icy sea; their quest for the lost Atlantis!
Lemmings were more to me than boreal rodents. They were
epic creatures; creatures of tradition; creatures of literature;
creatures to be mentioned with dragons, with leviathans, with
unicorns, with Pegasus!

And here, on snow-bound Shugliak, I had been capturing,
measuring, skinning, and dissecting lemmings!

Chapter XXX

"Mer' Klishmush! Mer' Klishmush!"

On the morning of December 24 I was setting lemming-traps in a snowdrift about a mile from the post. The thought of Christmas somehow had not crossed my mind that day.

I worked diligently, setting about two dozen traps. During the course of my morning's labor I must have moved a cubic rod of snow. But something was wrong. Some unfamiliar lack of interest made me listless and pensive. I had no soul for this endless lemming-catching. My movements were the movements of a snow-digging, trap-setting machine. Alas, I could not even keep my mind upon lemmings!

Was it that a far-away tinkling of sleigh-bells insisted upon making itself heard above the clickings and crunchings of my shovel? Was it that these lemming trails blurred and went lacy as I gazed, reshaping themselves into the hoofprints of six tiny reindeer? Was it that this sparkle of frost-rime kept flashing ropes of tinsel constantly before my eyes? These naked willow-twigs at my feet: were these only the hare-browsed willow-twigs I had been seeing day after day or were they, rather, neat, cozy evergreens waiting to be hung with spangles and strings of popcorn, and red and white candy canes?

Christmas Eve, of course!

Had not a hundred radio messages bade me look after Santa Claus and the reindeer and send them on their long journey in time? Had not Sam and Jack and I made solemn promise to one another that we would be at the post for "the holidays"? Had I not brought from the Southland a whole boxful of toys and trinkets for the Eskimo children? Lemming-trapping indeed, on such a day as this!

I gathered up knife and traps and shovel and started back to the post with a new lightness of foot. "Won't Sam and Jack be surprised," I mused, "when I burst in and remind them that it's Christmas-time?"

I was ready with a good jollying by the time I got to the outer storm-door. But what could I say to Sam when I found him at the kitchen table tying up big packages and scrawling *Merry Christmas* all over the paper with a red pencil? And what could I say to Jack, who was sitting on the floor upstairs in the midst of a confusion of string and cardboard boxes, old kegs, rumpled paper, scissors, hammer, and tacks? My only course, I clearly saw, was a direct and humble one. I beat the snow from my clothing, took off my hunting *kooletah*, picked the ice from my eyelashes and eyebrows, dug out of the corners and from under my cot all the knives, hatchets, beads, toys, shirts, underclothes, picture-books, and candy I intended to give to the Eskimos, hunted up some paper and string, and set to work.

There were, I had decided, five men, six women, six girls, and seven boys to whom I wished to give something. No Eskimo at the post was to be left out. I'd have presents for Sam and Jack, too, of course; but I was concerned for the moment chiefly with the Eskimos.

What did these Aiviliks know about the White Man's Christmas, after all? Would all this giving of gifts appear inane or sycophantic to them? Would they feel themselves obliged to make some return? And how could I avoid causing jealousy? I had little way of knowing which gifts would please certain individuals most; but I had decided long ago that I would give no gifts at all rather than give each Eskimo the same article. We would have a real Christmas, or no Christmas. Eleemosynary Christmases we would leave for the civilized folk back home.

Sam and Jack and I wrapped packages for hours. Jack wrote the names of the Eskimos on my parcels. I went so far as to add pictures of holly leaves and berries in water-colors. We made certain that none of the packages was too neatly done up, nor offensively clean. We did not wish to terrify our friends at the servants' house, nor to awe them with our day of days.

Night fell long before we had finished our task. When, at length, we stepped outside for a breath of air, we saw that bright red papers had been put up in the windows at the Mission. These cast a strange and lovely rosiness across the snow.

Just before supper three neat packages arrived by boy messenger from Father Thibert and Father Fafard. In two of these were fat doughnuts made from the Mission's precious store of sugar, flour, and lard. And in a small blue box with spring lid was a pencil for me. We were so pleased we were hilarious for a time, then suddenly quiet. I wondered by what strange process of reasoning I had come to believe it any more blessed to give than to receive. Then we scurried about, full of ideas. We filled a carton with chewing gum, sweet chocolate, tubes of tooth-paste, rolls of lint bandage, dry-skinned oranges, and one of my ten books. Jack and I raced over to the Mission with our gifts, extended the compliments of the season, and had some music on the organ. We couldn't remember the words of most of the carols, but we could remember the tunes.

For supper Sam and Jack and I had hare pie, a tin of as-

paragus, and some white French wine. So this was the desolate tundra; this the cruel Arctic! We sat back luxuriously in
our chairs as the plaintive strains of "Oh, Susanna, Don't You
Cry for Me!" from our two mouth organs mingled with the
friendly clouds of tobacco smoke, the heat from the red-hot
stove, and the crisp, icy drafts that shot down from the loft.

Shortly after supper Mary Ell came in, as usual, to wash
the dishes and sweep the floor. Jack and I were so busy making candy that we prolonged the dish-washing a good deal;
but we were all so happy it didn't matter how much work
there was to do. I painted a picture of a sprig of mistletoe,
cut it out with a knife, and tacked it above the kitchen door.
Then we chased Mary Ell round the room with a pair of owl
claws, grabbing her and kissing her whenever she got into the
danger-zone. Sam wondered what all the scuffle was about,
and when he appeared at the "office" door was a trifle
straight-faced; but when he saw the cut-out mistletoe there
wasn't much he could do but laugh and join in the frolic.

At an early hour in the evening visitors began to arrive.
First was old Angoti Marik, who greeted us one by one as he
came in, said *"Ikki"* to me, and sat upon a chair in the "parlor" with his feet little more than touching the floor.

The door of the outer porch slammed, the inner door
creaked open, and from the wheezing and puffing and swishing of skirts we knew that old Shoo Fly had arrived. Her
greeting was cordial, but her words and laugh reminded me
of the croaking of a raven. She shuffled to a chair and sank
upon it with a grunt and sigh. Her hair had, I dare say, just
undergone a rigorous combing, for it was tidier than usual.
Her sagging cheeks and broad forehead were decorated with
many tattooed lines.

For a time we gave ourselves over to small talk, agreeing

that the weather was cold, that the day had been fine, that foxes were plentiful this year, and that it was good to think of Christmas.

Angoti Marik (the old codger averred that he really was a Scotchman and not an Eskimo!) soon drifted into one of his outlandish stories. His face beamed as he squirmed, cleared his throat, and made quaint gestures.

"It happened on just such a night as this, *taipshimani*," [1] he began.

"Two girls with their old grandmother were camping in an *igloo* at the edge of a large lake. Another *igloo* stood near by, the *igloo* of the girls' uncle.

"The fat in their *koodilik*-lamps was so low, the wind so cold, and their supply of food so nearly gone that the girls went to their uncle's *igloo* to ask for help.

" 'Come, see how small our fire is; how cold our grandmother is; how little we have to eat!' said the girls. The uncle came out and together they all returned to the snowhouse where the grandmother was.

"As they approached, smoke began pouring from the gut window, the holes in the roof, and even the low doorway. 'How is this?' asked the uncle. 'You tell me you have no fire, yet there is much smoke?'

" 'We do not understand,' said the girls.

"Then flames belched from the doorway, and the poor old grandmother crawled out to die before their very eyes in the snow. The fire raged and black smoke poured forth until the entire snow-house, all that was in it, and the poor old grandmother were burned up. Not a thing remained."

We maintained a fitting silence during this dramatic recital; then, a certain incredulity getting the better of us, we

[1] Long ago.

said: "Scotch Tom, do you believe such a thing really happened? You know snow cannot burn."

"Oh, yes, of course it happened. Someone told me that it all happened. I myself know the man who told the story!"

Christmas Eve is no time for altercation, of course, but we pursued the matter further: "But, Tom, how could a fire start when there was nothing in the *igloo* to burn; and how could the snow burn even if there had been a fire?"

"It is strange," answered Tom. "But it happened, nevertheless. I know the man who told the story."

As if realizing that Angoti Marik's tale had no special connection with the holiday season, Shoo Fly began a long description of the olden times when at Christmas the crews of the whaling vessels gave themselves over to all sorts of revelry, making noise beyond belief, dancing like wild men, eating and drinking a great deal, and, all in all, upholding the traditions of the healthy Caucasian on a spree. This description seemed to me a justifiable hint that we were too staid on this evening of evenings. At any rate, Jack put gay records on the gramophone, we got out the accordion and harmonicas, and tried to liven things up a bit.

Embarrassment and reticence had not yet been dispelled, however. This I could tell from the solemnity of the faces about me. I decided it was up to me to break the ice somehow. I was the newcomer, the intruder, the only person in the room who could not speak Eskimo. Well, I would try.

I asked Sam to remind the Eskimos, once more, that it was the White Man's custom to behave in an unusual manner at the Christmas season. Shoo Fly nodded vigorously, but solemnly, in agreement, and added a word or two about the whalers.

With this as an introduction, I began cutting solo capers, doing my best at a sort of clog-dance, a song or two about

Dixie Land, and jumpings over chairs. Jack and I sang a duet we had practised three or four times, achieving fair harmony here and there. At this the Eskimos put their hands on the walls the better to feel the vibrations caused by the successful chords.

Scotch Tom and Shoo Fly laughed gaily. Shoo Fly beat her knees with her old hands, shouting that she knew I was a whaler now, after all, and an American or a Scotchman at that! All of us went almost insanely gay, so glad we were to be free of the stiffness that had possessed us. In a moment of inspiration I grabbed up our washbowl (it was filled with dirty water) and undertook an imitation of a waiter prancing round in a dining car. What had been laughter before developed into a storm of applause as the doors opened to let in more Eskimos and an icy breath of air from the outside. Sam and I had some trouble in explaining the White Man's trains and dining cars and waiters. Pictures in a magazine helped.

Shoo Fly was by this time well-nigh overcome. Her laughter had so worn itself out that she made but little sound, though tears ran across the blue tattooing of her cheeks and her whole body shook. Presently, before anyone had a chance to catch her, she fell from her chair, her head wagging from side to side. Two of the men set her upright once more.

I might have ceased with this *beau geste* from Shoo Fly. But once the Arctic Spirit of the Drama had taken me in hold it did not release me promptly. Hot from jumping about, I sat down on Sam's bed. Shouts came from the parlor: "*Ateeloo! Ateeloo! Piuyuk!*" [1] An encore of booted feet stamping, and a pounding of tables and chairs. Jack was chippering like a baboon and I never saw Sam so kittenish.

Heaven knows how I chanced to think of any further way

[1] "More! More! Good!"

of entertaining this enthusiastic audience, but I started again, this time using the doorway as a stage and presenting a series of imitations of animals known to live on Shugliak.

Snatching a tablecloth, white long since, I draped myself and ambled across the doorway on all fours, growling and swinging my head from side to side. This was *Nanook*, the Bear. With broom and coal shovel stuck up from my head, I gave them *Tooktoo*, the Caribou. Wriggling on the floor, I aped *Netchek*, the Seal. Holding two table knives to my upper teeth in imitation of tusks, I bellowed loudly. All the Eskimos in an instant shouted *"Aiviuk! Aiviuk!"* They knew a walrus when they saw one! They were willing to play this game, and to play it heartily! Now they fairly goaded me on to imitations of weasels and ravens and ducks and lemmings, and finally of animals that live in other lands, pictures of which they had seen in books. I began to regret that knowledge of Natural History was so extensive in this uncivilized place. I could not plead ignorance of all these creatures, for this would simply mark me as unskilled at my "trade." So on I went. Noah's Ark itself could hardly have been the scene of stranger capers than were presented in our kitchen that noisy evening. I wound up my performance with creeping downstairs on hands and knees: a lion in pursuit of a bush negro. The Eskimos had seen a picture of something of this sort and it was their idea of a dramatic situation. Applause was loud and long.

Now began a strange eulogy, a eulogy that was not, I fear, quite free of flattery. I was told that I was a good explorer; that I drew pictures well in the snow; that I drove the dog-team with unusual success for a beginner; that few White Men from the Outside learn the Eskimo language as readily as I had (true, I could remember the word *ikki*); and so on, and on, and on, until I was at the point of despair that I had not brought a whole shipload of presents to give to this kind people on the morrow.

At midnight we set off two great rockets. With a screech the shafts of fire leaped toward the low-hung stars, followed by their glowing trails of sparks. Dogs yelped, raced for cover, and broke into a chorus of howling. Sam and Jack and I ran about clapping everybody on the back and shouting "Merry Christmas! Merry Christmas!" And the Eskimos ran about too, clapping everybody on the back also, shouting "Mer' Klishmush! Mer' Klishmush!"

Then our visitors left us. We had supper again in the kitchen: cold hare pie, and a tin of lobster, and a lemon, thawed out on the stove.

.   .   .   .   .

Sunrise, at about ten o'clock, was a glorious effusion of pink and gold. At either side of the great day-star stood bright mock-suns; and upward from this tundra-king and his men-in-waiting shot narrow shafts of light, almost to the zenith. The upper sky was green as ice.

Hardly had we stirred up the fire and eaten our simple breakfast when Eskimos began to arrive, wishing us again a "Mer' Klishmush!" and carrying Sam's presents promptly away with them. Jack and I decided we would distribute our gifts later in the day. This, we foresaw, would prolong the merriment, and give us the coveted opportunity of watching the quaint folk opening their parcels.

At about half after one o'clock in the afternoon, just before sundown, all the Eskimos gathered at the servants' house for the visit of Santa Claus, and thither Jack and I went, struggling with our armfuls of gifts. We were neither bewhiskered nor becapped, and our new blue and orange sweater-jackets violated all Christmas tradition, but we felt as rosy-cheeked and red-coated as any Santa that ever lived. When we entered the small and crowded room we were greeted timidly. We threw down our burden with a shout and began handing out the packages. "Oo-why! Ee-yee!" shouted Angoti Marik as, seated in a dark corner on a crude bed, he unwrapped a checked lumberjack's shirt and small new hatchet. Shoo Fly was seated on the same bed, legs crossed under her, mending a boot. She wiped back her hair as she croaked recognition of her gift and untied it ceremoniously. All gazed attentively as she unfolded a white union suit and a tissue paper packet enclosing a black bead necklace and pale blue earrings.

There were pretty trinkets for all the girls; but for Mary Ell there was a sort of sweater-coat, with tasselled belt attached, of a shade of blue to be compared only with the most ethereal of heavens. At sight of this gift, and of the bright pink beads that went with it, Mary Ell blushed a deep brick-red color, let out a queer and agonized cry and, ducking low, scuttled into an adjoining room.

The many children had been so stuffed and gorged on tales of the White Man that they accepted everything with scarcely a quiver of eyelid. Soon balls were bouncing across the room, a mouth organ was whining from every corner, and tin motor-trucks were wheezing about, bumping into everybody's *komik*.

One little girl got a neat box of modelling clay. A cloud darkened her sky for a brief moment when she was told that this clay was not, under any circumstances, to be eaten.

Among the last of the gifts was a pink and white rattle that a friend had asked me to give, for him, to "some little Eskimo." Glancing round the room, I descried the black eyes of a baby, peering forth in solemn wonderment. Stark naked,

save for a little chain and amulet about its neck, it sat in its
mother's *kooletah*-hood, its hands clasping her brown neck.

I have a great respect for babies. In every book I read, babies appear to be the infallible judges of character, and I
walk trembling before them, realizing the chances are ten to
one they will begin to cry when they see me close at hand. I
was braver than usual, this time, however. I picked up the
package, sidled over to the little one, and unwrapped the rattle. No scream rose, no rains descended, nor floods came; the
little thing, patting my chin with one hand, grasped the gay
rattle in the other and shook it lustily as if all its eighteen-
month career had been a rattle-shaking career. Its short-
lashed, narrow-lidded eyes worked themselves into an unmistakable grin. What a victory, especially for a blue and orange
Santa Claus!

As we were preparing to leave, we noticed two men, sitting by a window, silently and wistfully watching the crowd.
They were Sheeloo and Tapatai.

"Haven't they received any presents, Jack?" I asked.

"The two men just comes in from Salmon Pond, Doc,"
Jack answered. "These people doesn't know anything about
Christmas way over at Salmon Pond. They just comes in to
trade their foxes."

We told the men that we had some hatchets for them at
the other house because it was our Christmas-time, and that
they were to come with us. With a terrible shout that
drowned all other racket for the instant, they sprang up gaily
and followed us crying "Ee-yee! Ee-yee!"

In the dusk the children played outdoors. Mouth-organs,
marbles, candy, tin motor-trucks, modelling clay—all had
been forgotten. Red and blue and white rubber balls lay
about for the dogs to sniff. The boys had returned to their
favorite game of sliding down drifts and rolling about on tin

cans and bottles; and the girls were chasing one another hither and yon, laughing and screaming like creatures beside themselves.

In the servants' house one little boy remained. He had found that by holding a mouth-organ close to the lips of a snoring man the strangest of sounds could be produced. He alone was playing with his civilized present rather than frolicking in the snow.

CHAPTER XXXI

Blue Fox Tracks

WE HAD many a discussion, there at the post, about the creature called the blue fox. You recall that these "blues" were rare on Shugliak. They brought considerably more in trade than the "whites." The Eskimos were always in hope there would be a "blue" in one of their traps.

Whenever I held forth on "blue" foxes I tried to make it clear that naturalists considered them only a color-phase of the common *Teregeneuk*. They were very different in color, to be sure, but they were really only white foxes with what, for the want of a better word, we called an *abnormal* coloration. These statements that belittled the much-talked-of "blues" met with strong opposition.

"But the blue fox is very different," my friends would say. "Different in color always, both in summer and in winter! Longer hair on the feet! Finer fur!"

And what could I offer, after all, by way of justifying my position? The books I had were only books, and the North Country has a fine scorn for anything that has been printed about the North Country. I couldn't breed foxes right there in front of everybody proving to them that occasional "blues" are born with litters of "whites." Part of the trouble was traceable, of course, to the mere sound of my words. My saying that a blue fox was really only a white fox was a good deal like saying that a black raven was not really a *black* raven, but a white one.

Finally I gave in. My friends were always being courteous to me in subtle ways, so why shouldn't I extend subtle courtesies in return? I gave in, but I did so with a twinkle in my eye. Coming in from my trap-line one day I set the post agog by announcing that I had seen, at last, the trail of a blue fox.

Everyone was a bit taken aback, surprised that I had taken the right stand regarding this rare beast.

"But how did you know it was the trail of a blue fox?" they asked. "Were the tracks larger, or shaped differently?"

"No the tracks weren't larger," I answered. "In fact, they were precisely the same as any fox track, save that they were *blue*."

Silence for a moment. I did not smile. Sam and Jack, "catching on" immediately, did not smile. Nobody smiled.

"So!" the Eskimos said. "The *tracks* were blue!" They looked at one another, nodding courteously, blinking as if a trifle bewildered but not in the least as if skeptical. "The Doctor says he saw *blue* fox tracks today. The tracks themselves were blue. Must be so; must be *blue* tracks if the Doctor saw them!"

I thought it rather a good joke. It was the first joke of the kind I had perpetrated. I was happy to be feeling like joking, for this feeling was proof of friendliness between us. I was, however, a little surprised that no one laughed.

Three days later John Ell came to me. His manner was more formal than usual. It was as if we had had an appointment. I could see that something was on his mind.

"What is it, John?" I asked.

"Docta, she fin' blue fox clacksh, so?" he somehow made out to say. His eyes had a certain moistness, I thought.

Yes, I had seen the tracks of a blue fox, a few days ago. I remembered them distinctly.

"Docta, she wanna she blue fox clacksh *oobloomi?*" [1]

Why, yes, of course I'd like to see some more blue fox tracks. I was always interested in these phenomena, to be sure.

I went with John. Not far from the house we were joined by a little company of Eskimos, all of them solemn of face.

We walked to the top of the hill back of the Mission. We came to the trail of a fox.

There in the snow was as neat a line of fox-footprints as you ever saw, real fox-footprints too, only in the bottom of each footprint had been placed a little round piece of blue tissue paper cut from the covering of a bat of absorbent cotton.

Now I ask you, is the blue fox only a color-phase of the white fox? The very idea!

[1] Today.

CHAPTER XXXII

The Dead of Winter

JANUARY and February of the year 1930 were low months for me. I trapped fox after fox, weasel after weasel, and bushel-basketfuls of lemmings. But in visiting my trap-lines I plodded through the snow methodically, wondering what it was all about. I made a considerable series of water-color paintings, too; but I painted laboriously, growling at the water that was everlastingly freezing in the cup; at the brushes that went dry in the middle of a stroke, scattering minute crystals over the paper; at the dim and varying light.

My January-February diaries are a mess. I will never show them to you. Were you to read them you would say to yourself "Aha! I thought so!" and then laugh rudely.

Well, perhaps I *was* a little lonely now and then. Frozen noses are so wearisome. They never get well. Ingrown toenails are positively almost a pain in the neck. Clark's Tinned Rations are fit only for the dogs. Lemming-skinning is the lowest known form of drudgery. Reading is so deadly futile. Radio announcers are so damned smarty, cackling away about snow and ice and polar bears and Santy Claus and all that *muck*. Why don't the folks back home send messages that aren't so cheerful? What's the big point in laughing all the time, after all? These bird pictures, just look at them: paper smeared all up with stuff called color.

I had not, you see, mastered the art of not-thinking. Therein lay my January-February difficulty.

It was faintly amusing, I remember, to hear Sam saying, in his patient, philosophic way: "Well, Doc, winter is coming at last." Just as if we hadn't already had four months of it. The thermometer at the corner of the house registered fifty some below at the time, and the water in my tin wash-basin was solid as a brick.

Dry cold, I suppose; the sort (according to various weather conversations I have sat in on) that you do not notice. And my face was frozen—dry frozen, I suppose—the sort of frozen face you do not notice. *Dry rot,* if you ask me, all this talk of dry cold that you don't notice.

The Eskimos were looking at me dubiously these days. I thought they were concerned with my mental state and its possibilities. But it was my nose that worried them. *"Imacha kingak nowk!"* Angoti Marik would say, which, being translated is: "No nose, maybe!" The dropping off of noses, fingers, and toes is not unheard of in this dry-cold Arctic. The thought of going home *sans nez* was, *sans doute,* a most unpleasant thought.

Angoti Marik's harangues on freezing were chilling to my blood. He would begin by telling of this person or that person who had frozen his toes. Then it was this person or that person who had frozen his feet or his ankles. Then it was somebody who had frozen his knees or his hips or his hands or his wrists. Angoti Marik's gestures were unfailingly to-the-point. He would discuss a frozen toe in great detail then, with a flick of his hand, wave the poor, gangrenous, worn-out member off into oblivion and begin on a frozen ear. By the time his discussions were over he had put his mobile hands on practically every part of his body and waved away all the toes, fingers, nose, chin and ears. I had the feeling, as I sat there listening and watching, that there was nothing much in the way of human beings left in the world but a lot of wretched, armless, legless, more or less headless, more or less insensate trunks. Really you must ask old Angoti to discourse on freezing for you some day. After you have heard him you will be glad you live in the South Country.

. . . . .

Tooahtak *komatikked* to the post to tell us of five hunters who had drifted off on the ice beyond Munnimunnek Point, and drifted back. They had been gone six days. They had seen and shot a great many *oogjook*-seals. But one of the men was almost dead from exposure. His feet had been frozen badly and his toes were dropping off. Sam immediately sent some medicine and bandages down to Munnimunnek.

. . . . .

We had an art exhibit one evening. Angoti Marik and Shoo Fly and many another leading-light were there. All of my water-color paintings of birds and mammals were spread out on the table. It was quite an affair, what with the big dish of gum-drops also in the room. Approbation was comically noisy. Groan followed groan; grunt followed grunt. "Why-ee! Why-ee!" could be heard all over the room. "*Wah kud-lunga!*" [1] said some. This was a great compliment. Finally Shoo Fly, who had been accustomed to a place of prominence in affairs social ever since the old days at Repulse Bay, but who had not, I dare say, ever before found herself in the rôle of art critic, delivered her ultimatum. "You are not human. You are more like a camera!"

Refusing to think of this as anything but a kindly remark (though cameras are, to be sure, the least human of apparatuses), I told Shoo Fly that I had made all the pictures myself with brushes and color and water.

1 "Wonderful to tell!"; "Marvellous!"; about an equivalent of Virgil's "Mirabile dictu!"

There was a further inspection. These, then, were not magazine pictures quite. Finger-nails were gently dug into the paper in an attempt to lift and rumple the plumage of the birds (was ever a compliment more deft?), and someone said *"Ikki!"* [1] when he put his hand against my drawing of a frozen lake. Oh, I tell you these Eskimos know their blarney. Be not mistaken on that score!

KIGAVIK: THE WHITE GYRFALCON

But Shoo Fly said (the skeptic!): "I want to see you doing one."

The next time I was painting I sent for Shoo Fly. I had already drawn in a ridge, a lake, and part of a white gyrfalcon. In Shoo Fly's presence I stroked in some brown rocks, some yellow sky and some gray barring on the plumage. Shoo Fly picked the picture up, her face beaming. And before I could detain her she had swept her hand across the paper in a final gesture of belief and approval. The paint was far from dry. Rocks and sky and gyrfalcon were now a hopeless smear. But Shoo Fly was leaving the room in a magnificence of swishing. She knew at last that my pictures had not been cut from

[1] "Cold!"

a magazine, developed in a dark room, or powwowed out of the ether.

· · · · ·

One of the bitches at the post kept following the trap-lines. One day she was caught in a fox-trap and her wailing and moaning could be heard from the head of the bay. In response all the dogs howled and howled. The noise got on my nerves. I suggested to Jack that we find the bitch and release her.

But Jack said: "The people here isn't kind-hearted that way, Doc. They just lets the dogs suffer!"

And that was that. The point is, I learned, that if you release the dogs they form the habit of following the trap-lines day after day. If you leave them in the traps awhile they learn to stay at home.

· · · · ·

We had a map-drawing session. John Ell and Tommy Bluce were to make me some sketches recording the Aivilik concept of our island. Tommy's drawing interested me greatly. It was not a continuous outline. It was a sort of delicate fringe of short strokes like the conventional representation of mountains you have seen on maps. John's and Tommy's maps agreed in being strikingly different in many particulars from the published charts.

· · · · ·

Munnapik came in to trade telling us of a bear-den he had found along one of the *noovoodlik*-hills. The mother bear had been sound asleep; but the cubs, which were about eight inches long, were suckling their mother greedily. Munnapik brought with him several bagfuls of *ichalook*-trout from one of the big lakes. The Eskimo name for this lake was the same as that for a common variety of seaweed, the reason being that the *ichalook* were so numerous beneath the ice that they had the appearance of seaweed.

. . . . .

The women sewed outer soles of bear-skin on my *komik*. These soft pads were warm and comfortable, though I felt like an animated scrubbing brush whenever I walked through the house, and the dogs showed entirely too violent an interest in my feet when I walked near them. Dogs are easily excited by the odor of *Nanook*.

One morning I set out for my traps wearing my new bear-skin soles. Two teams of dogs that had just come in from Cape Low followed me. This was strictly against the rules. I shouted at them, but they did not go back. They stopped, sat down, and waited for me to resume my walking. Then they rose and followed. I shouted all sorts of threats at them but they were persistent. I threw my trapping hatchet at the nearest one, hitting him on the flank. He ran back a way, then stopped. When I chased the dogs they ran, of course; but they stopped when I stopped and followed when I walked forward. It was not at all pleasant, this wolfish stealthiness. Finally it dawned on me that I should have to deal with them in Eskimo.

I shouted out the only fitting word I could remember. It was *"Ahtay!"*, just the simple word *"Ahtay!"*,[1] the word Little Peter used when he asked me to "play animal" with him. But I shouted it in my roughest voice. And, believe me or not, the dogs all turned tail and slunk back to camp as if they had seen an evil spirit.

. . . . .

[1] "Now!"

During February everybody brought me embryo *netchek-*seals, small, cream-colored, woolly creatures taken from the gravid females that had been killed. The women made mitts and bags from the skins of these unborn *netchek.*

.　　.　　.　　.　　.

One day Kooshooak and I were out with the *komatik.* Kooshooak told the dogs to "find animal." They ran at breakneck speed straight across a lake and came to a halt at a dead lemming on a rock. They had scented that little carcass probably three miles away.

.　　.　　.　　.　　.

Khagak paid us a visit, bringing his accordion. Khagak knew only four English words. These he always said all at once. And it is doubtful that he knew they were four words, for he always said them as if he thought them two: "Mygodboy accordion! Mygodboy accordion!" There was a little separation between the first three and the last words, for Khagak knew definitely what the accordion was, and played it with gusto. He made me an old-time caribouskin drum, a musical instrument that was once widely used among the Aiviliks.

The caribouskin drum has all but disappeared from Shugliak's *tupek* and *igloo.* Harmonicas, gramophones and accordions have come to take their place. And the Aiviliks (at least some of them) sing such civilized melodies as *Eema, Peetohungitook Bah Nannash,*[1] Annie Laurie, and Nearer, My God, to Thee.

.　　.　　.　　.　　.

Jack and I showed some of the Eskimos the photograph of a statue representing a smiling man. We explained, as best

[1] Yes, We Do Not Have Any Bananas.

we could, that the man was not a real man but only a stone image, something in the nature of a *tooghak*-carving. "That is better," said the widow Kuklik; "a man of that sort smiles all the time!"

．　　．　　．　　．　　．

A snowy owl specimen arrived from Cape Low. It was a good specimen, save that one of its feet had been cut off in a trap and a new foot from another owl sewn on. The objection I had to the new foot was not that it was from another owl, but that, like the intact limb, it was a *right* foot, and it had been sewn on in such a manner as to give that appendage two heel joints. But what is an owl's heel-joint more or an owl's heel-joint less to an Eskimo?

Chapter XXXIII

Viscount Grey

ONE morning in February a half-grown collared lemming was making his way peacefully through a burrow he had dug through the snow not far from the servants' house at the post.

All at once there was a loud sound in front of him and a loud sound back of him. He stopped. Light dazzled him. Bewildered, he ran forward, to find himself in a half-built *igloo*. He was terrified. But when he tried to run back into his burrow he found himself held firmly in a big caribouskin mitt.

A shy, black-eyed, tittering girl brought this lemming to my workroom alive. I paid a paper sack full of red and white candy for him. I decided to call him Viscount Grey.

Viscount Grey was, from the first instant of our acquaintance, both friendly and confiding. He nestled in my hand, coughed and chuckled when I touched the tips of his dainty vibrissæ, ate bits of cooky and raisins from my fingers three minutes after he had smelled of me thoroughly and nibbled at me with his sharp teeth, and promptly declared himself half-owner of my worktable. He wanted to feel me with his whiskers, to pretend panic, to race off in mock terror to return for more excitement. But I had work to do. I set him on the table, three feet away from my own workground and told him to mind his own business.

Over the table he raced, sniffing at everything, touching everything with his "feelers," biting at everything with his sharp incisors. He licked the wire-cutters, tried to run up the side of the glass kerosene lamp, scuttled through the powdered arsenic, sneezed, chewed at the bark of a willow-twig, splintered a match, fussed with some cotton wool, balanced on the end of a file, peered far out over the table edge, stood on his hind legs and surveyed our workshop, came over to taste a tiny drop of perspiration from my finger-tip, and clattered into the tool box, scratching with his big, double-tipped, clumsy digging claws. Eventually, forgotten for the moment, he began an examination of diaries and note-books, riveted his attention upon one sheet of data concerning the Herring Gull, and proceeded to chew, leaving a lace-fringed hole in the paper and many hiatuses in the annotations. Even a Sherlock Holmes would have thought twice before trying to reassemble the tiny bits of paper that Viscount Grey scattered about, kicked aside, or swallowed before he was thrust into a tin can, where he fretted considerably.

He soon became the pet of our community. Little Peter screamed with delight at feeding him raisins. Sam played with him by the hour. The missionaries smiled at him and stroked his soft back. We took motion pictures of him as he ate, preened himself, went to sleep, and dug holes in the snow. He, of all creatures, once made John Ell actually jump in a spasm of fright. John had hunted every fierce beast in the Arctic. You may recall his improvising a harpoon from snow-knife and boat-hook for a hand-to-hand battle with a bear. But all the same Viscount Grey made John Ell jump. Entering the room where John sat conversing, I put Viscount Grey quietly on his knee. The fearless hunter recoiled, jerked a little, and smiled; and fresh drops of perspiration glistened on his forehead!

Viscount Grey! So much a part of our family did he become that he ate at the table with us, wandering from plate-edge to plate-edge, eating what he chose and showing a marked preference for cookies and jams. His favorite retreat was the inside of an all but empty coffee cup where he licked up what remained before tucking himself together for a nap. He might have lived with us all winter; but I was away from the post so much of the time that I decided to liberate him. I took him out to a deep drift, put him down at the entrance to a burrow and sealed him in.

What tellings of tales through touchings of whiskers there would be in the blue twilight under the snow!

Chapter XXXIV

The March-"Month" on Shugliak: *Netchialoot*

THE *netchek*-seal is by far the commonest seal found in the waters surrounding Shugliak. From the Eskimo standpoint the *netchek* is, therefore, the most important marine mammal of the region, for it is the source of most of the sealskin, blubber, and dog-food needed in their daily lives. *Netchialoot* is the season when the female *netchek* give birth to their young.

Netchialoot is a season of sparkling days. Days that are ideal for *komatik* travel, save that the glare of sun and snow are wearisome. Water trickles down the sides of boulders at noontide, freezing into thin ice-ripples by evening. Ptarmigan bask in nooks along the ridges, their eyes almost shut. Bright-capped *shukshighiuk*[1] swing about from willow-clump to willow-clump, calling wheezily. And the mother *netchek* guard their soft-eyed, cream-colored babies in sun-bathed ice-nests far out in the frozen bay.

Netchialoot is a season of wild winds that bury the fox-traps, cover or blow away the *komatik*-trails and gouge out the drifts, exposing thousands of lemming burrows to the cold air. It is a season of snow-blindness among the Eskimo children. It is a season of yawning and stretching *angenuk-nanook*, that lick and nuzzle their cubs, lifting them from their dens and leading them slowly out to the seal-dotted *sheenah*.

[1] Redpolls, *Acanthis linaria* and *A. hornemanni*.

224

We were busy during the season called *Netchialoot*. Kooshooak brought in a coal-bag full of *pitseolak* specimens from
the Bear Island floe. These *pitseolak* are known among ornithologists as Mandt's guillemots.[1] They are pretty, red-
footed diving-birds that feed on the pink crustaceans Sam
and Jack called "sea lice." Kooshooak had not cared for his
"specimens" very well. He had not shaken the coal-dust out
of the bag and the white plumage was badly soiled. He had
broken the leg- and wing-bones in the belief that such treat-
ment of his quarry would assure him soundness of body in
his old age. He had thrown the birds together in the bag,
then used the bag as a sort of cushion on his *komatik*. The
frozen heads and necks and legs and wings were broken off
and mixed up so badly that it was no small task merely putting the pieces together before thawing them out for skinning. Thawing required about two days, for the thick plumage that had kept the birds comfortable in the cold sea also
kept the carcasses frozen for a surprisingly long time in the
house. Kooshooak told me the *pitseolak* were so numerous at
the floe that he had killed seventeen of them with one dis-
charge of my shot-gun.[2]

· · · · · ·

Muckik amused us and rather made our hair stand on end
with talk about a bear-den he had found. In coming to the
post from Native Point he had stopped at his usual camping-
ground near a low ridge. He had had no difficulty in locating
the walls of the *igloo* he had built only a week or so before.
But the inside of the ruined *igloo* was a yawning cavern and
from the cavern three trails led out across the snow: the
trails of a mother bear and two cubs. Muckik had lit his

[1] *Cepphus grylle mandti.*
[2] I loaned my shot-guns frequently so as to get good specimens. *Pitseolak* shot
with a rifle were worthless to me.

koodilik-lamp and eaten his fish and drunk his tea probably not more than six feet above the sleeping bear. And the dogs had curled up and dreamed their savage dreams all but nestled against the back of their most thrilling foe!

How very exciting to waken in your *igloo* with a heaving and growling of wakening bear beneath you! Or to watch the smooth top of the drift on which you are standing suddenly crack and lift up and let forth an *angenuk-nanook* and her cubs! Tundra magic of another order, this shaking of bears and bear-cubs out of snow-drifts!

. . . . .

Eevaloo told us of a wolf that had been digging out and springing his fox-traps. The clever animal apparently clawed the snow from under the traps, stealing the bait and snapping the steel jaws shut by knocking the traps about in the snow.

. . . . .

During stormy periods I packed the specimens I had collected and prepared during late summer. To my surprise I found that many of the larger bird-skins that I had stored along the edge of the room had frozen solid to the wall. I had to use care in thawing out the ice and in drying the plumage thoroughly.

. . . . .

We had a disagreeable four-day warm spell. The drifts went soggy everywhere, the crusts rotten and infirm. Water gathered on the lakes, adding a new fierceness to the brilliance and glare, and a blueness vibrant as the blueness of a morpho butterfly. The roof of our house leaked badly; or, to put the matter somewhat more accurately, the ceiling upstairs dripped a great deal of water. The roof was solid and sound enough; but a layer of frost had gathered on the upper surface of the beaver-board lining of the rooms, and this frost melted rapidly, letting a veritable torrent down. I did the best I could to cover specimens and books and paintings, but everything got more or less wet. I ran round the post gathering bottles and cans and stood about fifty of these at strategic points on the floor. The sound of dripping was musical enough, if you cared to think of it that way. I was emptying cans and bottles practically all the time, or wiping up the floor with old blankets and clothing. Finally we took some of the stripping off and let the water out all at once.

. . . . .

Father Fafard and Father Thibert showed me an interesting collection of implements that had been made and used by the extinct Shugliamiut: little ivory and bone spools or bobbins for *eevaloo*-thread; dart-heads; needles made from slender tarsal-bones of birds; decorative bits that may have been used as amulets. Most of these relics had, I understood, been found about the Native Point encampment. Among them was a strange pair of snow-glasses made more or less in the style of what we call spectacles: the eye-pieces thin rectangles of bone with narrow horizontal slits. These eye-pieces reduced the amount of light that could reach the retina, and probably in large measure prevented snow-blindness. Shugliak's Eskimos of today wear snow-glasses that are imported from the Outside.

. . . . .

I had an interesting experience with some ptarmigan. In following my trap-line one morning I came upon a large flock of the birds feeding on willow buds. They ran about nervously and lifted their red combs when I shouted at them.

On the way home that evening I noticed that a whole section of my morning's trail had disappeared. This was puzzling, for there had been no wind and therefore no drifting of snow. On coming up I found to my astonishment that over sixty *consecutive* footprints were neatly filled each with the plump body of a roosting ptarmigan! Some of the birds had already started to doze. As I walked along they did not budge, but opened their black eyes and looked at me fearlessly from their snug beds.

CHAPTER XXXV

A Baby *Nanook*

WE WERE seated as usual, that evening late in March, in our
"parlor" at the post, content to leave the Outer World to the
mercies of the wind. In the large chair near the radio sat
Sam, a cigarette between his fingers. On the bench lay Jack.
Two Okomiut men, Sheeloo and Tapatai, who had come in
late in the afternoon from the *noovoodlik*-hills, were perched
on chairs near the table. I must have been seated in a chair
myself.

Since foxes and fox-tracks, dogs and dog-food, weather,
sheenah, and the day's small doings had been discussed, re-
discussed and fully commented upon by all, the tide of conversation was at extreme ebb. Sam looked steadily at the
kerosene lamp. Jack gazed at the ceiling. I riveted my attention upon the photograph of pleasant Edward, Prince of
Wales, that was fastened into an old mirror frame. We three
white men had worked together, eaten together, and sat together thus of evenings so frequently that we knew one another's thoughts fairly well. Not one of us, however, had a
ghost of an idea as to what the silent Eskimos were thinking.

Had any of us remembered, at that awkward nadir of the
evening, that an Eskimo will undergo nearly any mental or
mento-moral hardship for the sake of creating a dramatic
situation, we might have known that the dead calm portended
a storm. We might have remembered Munnapik, who always
understated his catch of foxes merely for the satisfaction of

watching the change on our faces as he drew from his *komatik* five times as many pelts as he said he had. Or Noah, who would answer "*Ah-mai*" [1] to all our questions until he saw we were at the end of our endurance. Or Kayakjuak, most annoying of dramatists, who, announcing that he had lost some valuable article such as a shot-gun, would lead us into a long discussion of or abusive tirade against carelessness, only to assure us that, of course, he had "found the article again long ago."

Suddenly Sheeloo moved, shifted his pipe, and cleared his throat. The Eskimos were going. We restrained ourselves from shouting out "Good night" in unison.

No, the Eskimos were not going.

"*Atiuktamik netchukpooghut umaiumik!*" said Sheeloo, a sly smile on his face. "We brought a live baby bear with us!"

The sentence had the effect of a fire siren's shriek. Backs stiffened, faces turned, jaws dropped. It took us a full instant to perceive that Sheeloo probably meant what he had said.

"Well, where is it now?" we all but shouted.

"It is outside. We brought it on our *komatik*."

Of course it had been brought on a *komatik*. Of course it was outside. But where? Confound this Sheeloo! he was at his dramatics.

"Did you bring it in a box?" We were on our feet.

"We do not think it was in a box. We brought it in something more like a sack."

"Well!" we shouted, taking it for granted the beast was in a sack or crate of some sort. "Where did you leave your *komatik*?" We were rushing for the door. "Don't you know the dogs will get it?"

At mention of the dogs the Eskimos rose in heavy haste. "We forgot about the dogs," they said.

[1] "I don't know."

Without proper clothing, we ran into the night. A faint moon helped us on our way across drifts, among boulders. A dark, canine form was making its way round the corner of the store. Three more forms were following, in close pursuit. I thought I caught a glimpse of a light-colored object in the jaws of one. Was one of the forms shaking something? Was another choking down a gluttonous mouthful?

We located the *komatik* at last. Of course the cub was gone. We hunted for some time before finding the tattered remains of the coal-bag in which it had been brought to the post. It was useless to look further. Already the dogs were curling up in the snow, licking their paws.

Next morning I searched in vain for a claw, a wisp of hair, or a bone. The unfortunate creature had been utterly annihilated.

I found myself wondering why, in a world where justice of some sort is to be looked for, the Okomiut race of the Innuit had not perished ten thousand years ago!

CHAPTER XXXVI

Spring?

THE season that is called *Netchialoot* is followed by the season called *Terriglullioot:* the time when the young *oogjook*-seals are born. Young *oogjook* are gray-black, with an indefinite marbling or mottling on the head and back. Their skin is used for mitts and for decorative strips on the tops of *komik*. The season called *Terriglullioot* is about the equivalent of our month April.

Terriglullioot began with a sleety rain: a miserable day. The rain finally ceased, and a gale blew up. Snow fell in clouds. Windows at the post crashed in. A canoe flopped into the water at the Native Point *sheenah* and sank. The face of the tundra changed. But why speak of snow or wind or blizzard any more? Such talk is tiresome by this time.

.   .   .   .   .

One of the pups made the mortal mistake of swallowing the breast-bone of a ptarmigan. The bone lodged in the unfortunate animal's intestine and obstructed the passage of food. Sam said there was nothing we could do. But the Eskimos continued to think that the *kingmiatsuk* [1] might recover. Part of the time it rested easily, to all appearances; and then it would writhe and moan pitiably and I would be at the point of attempting an operation. Sam finally sent word to John Ell that the pup would have to be killed. Through

[1] Little dog; hence, puppy.

the snow-laden wind came little Ookpik, a stout thong in her hand. This she slipped over the pup's small head, and swung the jerking creature off through the storm. She hung it to one of the projecting timbers of a high rack near the servants' house, where caribou meat was cached out of reach of the dogs. Here the thin body swung and bumped in the wind until its struggling ceased. And here it continued to swing and bump until the long winter was done.

    .   .   .   .   .

When the gale had spent itself Jack and I decided we would *komatik* to the head of the bay to set traps and locate some fox-dens. Nowadays we wore snow-glasses whenever we were out-of-doors. I had had several touches of snow-blindness and did not want any more of it.

On the way home from our trapping-ground I asked Jack to let me drive the dogs. I got on fairly well for a time but began to have troubles. I think I was too eager to impress my young friend with my prowess as a teamster. We somehow got off the trail into deep, soft snow. The dogs were bewildered by my shouting, by this whip that flopped here and there among them, by this tall man that ran up pulling at traces and kicking and barking out vain threats of one sort or another. Jack was all for helping me, but I told him to stay on the sledge and enjoy himself.

Finally we got back on the trail. The dogs were plainly relieved and happy. In my gaiety I cracked the whip with renewed vigor. But I decided to quit monkey-shining when I heard that wicked tip snapping somewhere back of me and turned to see Jack laughing uproariously. I had cracked a cigarette out of his mouth, missing his face by an inch.

    .   .   .   .   .

On April 5, John Ell and I took a long trip to Itiuachuk. We got back at about eight o'clock in the evening, cold and tired. I glanced at the schedule of radio programs and conquered a pang: we were to hear nothing from Station KDKA until the following Saturday evening. I guess I must have been a little eager for contact with the Outside World.

I forget just what I was doing, there in the kitchen: taking off my *komik*, probably, or combing my hair at the little mirror near the table. Jack was in the parlor, seated near the radio.

"Doc! Quick, Doc! Somebody's talking to you, Doc!" Jack shouted.

It was Louie Kaufmann, at KDKA, back in Pittsburgh. Messages were coming through. Reception was perfect. Jack and I sat there spellbound, completely spellbound. The special broadcast that was listed in my note-book for April 12, was on. I just chanced to be there to hear it, there at the post rather than twenty miles away, over beyond Itiuachuk.

There was much of gaiety and friendliness about it all. I could scarcely believe my ears when I learned that we were to hear my mother playing the piano, my Alma Mater's orchestra performing just for us, and the boys from my college fraternity singing some of the Beta songs I loved so well. Jack and I sat there, scarcely moving, scarcely saying a word, for almost three hours. It was a soul-stirring, never to be forgotten experience. The Eskimos came in one by one and sat about. They said little, for they could see we were deeply interested—more interested than we had ever been before in the radio.

When the fraternity songs were being sung I hummed the tunes by way of expressing my friendship for those young men who were singing for me in Pittsburgh.

"Why did 'the Doctor' sing with the men who were singing?" John Ell asked Jack, talking, of course, in Eskimo.

"The men were 'the Doctor's' brothers," Jack replied.

"Mush' be ver' many family!" John said to me, smiling with a very likable expression on his face, an expression close enough to a proud expression to make me a little proud myself. It was a little as if, now that we had all become such good friends, John's own brothers had been singing to us, his brothers, or Jack's brothers, or Muckik's brothers, as well as mine. I couldn't help feeling very happy. A little sad too, a little lonely: but first of all happy. Perhaps some of this mysterious tundra-happiness was becoming mine—happiness along with ingrown toe-nails; along with arsenic poisoning under finger-nails; along with sunburn and snow-blindness; and frozen nose.

. . . . .

I spent a good deal of time skinning out caribou heads and feet and in sewing together and preserving the big hides. On the inner surface of the hides, especially in the region of the rump, I found quantities of encysted bot-fly larvæ, small white grubs that would have lain under the skin of the living caribou and emerged for pupation sometime in summer I suppose. I remembered seeing some of the adult flies somewhere. They were big, noisy brutes with a nasty bite.

. . . . .

Jack and I had a comical session in taking photographs of some of the women. The men loved having their photographs taken, or at least they acted as if they loved it; but most of the women acted as if the camera struck terror to their hearts. We would get them into some desired attitude only to have them bow low or shrink down or scuttle away laughing that slushy laugh of theirs. Shoo Fly was matter-of-fact about the whole camera business. But the other women said: "We do not like the way the black box winks its eye."

. . . . .

The Okomiut were successful in their walrus hunting at the Munnimunnek floe. On one trip they got a huge bull, several cows, and a new-born calf. This strange baby *aiviuk* I was to have the pleasure of sketching in the field. What a queer *old* face for an infant: that wrinkled blob of a nose; those little eyes! And the mother: what a weird monster she was, with her slender tusks; the patches of pale pink about her mouth; the tough, thick whiskers; the oddly shaped pupils of her green and red-brown eyes!

. . . . .

On the ninth day of April we saw a snow bunting. Little black and white *Amauligak,*[1] returned at last from the South! Warm days must be at hand! He perched in the shelter of the half-buried motor-boats. His chirping brought happiness and hope of some sort to all of us. Even the Eskimos turned out to look at him. But the next day he was gone. Perhaps the dogs had got him.

As I gazed at the solemn whiteness that lay about me I felt that Springtime would never come.

[1] Snowflake or Snow Bunting, *Plectrophenax nivalis.*

Chapter XXXVII

Some Aivilik Tribal Stories

IF EVER you spend the winter on Shugliak you will hear some of the quaint stories that are told among the Aivilikmiut. You may have trouble in translating these stories, even though the words are spoken with more than usual distinctness. You will probably have to ask your friends to repeat many of the words. But any trouble you may have in translation will, I predict, be as nothing when compared with that of deciding just what the stories mean.

Story-telling among the Innuit is a social event. The Shugliak Aiviliks have no village hall or public meeting-house, as do certain other Eskimo tribes. But when stories are to be told several families gather in one of the larger *tupek*, the little audience arranging itself as comfortably as it can. The men smoke. The women nurse their *nootarak* or chew-down the edges of skins. The children sit in rows on the bear-skins or sleeping-bags.

Stories are told deliberately, in a dignified voice. Good Innuit is used; that is to say, the sentence structure and vocabulary are good; there is no "slang"; and the words are carefully enunciated. The children are learning these stories as they hear them; and they must learn them correctly.

The story-teller (usually one of the *angekok* of the tribe) uses all sorts of gestures as he proceeds. He moves his hands about, wags his head, and makes big eyes and narrow eyes and opens his mouth wide as the narrative warrants such grimaces. Either because they are completely transported, or because they are hoping to tell the stories in the same way themselves some day, the children imitate the *angekok's* gestures and facial expressions. You will smile when you see a row of youngsters round-eyed while an owl is talking; sticking their heads up out of the water while a *netchek* is singing; or peering under a rock while a weasel is asking a lemming to come out and chat with him for a moment.

The story-teller is an actor. When his hunter-hero is creeping up on a sleeping walrus every person in the *tupek* has a wild gleam in his eye, and you will decide that an actual harpooning of some sort must be going on before your very eyes. You will half-expect to see thick blood on the floor in a moment or two, or to hear the hideous bellowing of the dying *aiviuk*.

But you must hear some of these stories. There is one that we will entitle "The Raven and the Fox." Giving this title may not be in keeping with the spirit of the *tupek*, for I am not sure that the Aiviliks ever give titles to their tales. At any rate, here is an Aivilik story about a raven and a fox:

· · · · ·

In a cold land where there was no sun lived Toolooghak, the Raven, and Teregeneuk, the Fox. Toolooghak and Teregeneuk were very similar, save that Toolooghak was black and Teregeneuk was white. Both had four feet. Both ran about on the ground. There was so much snow everywhere and the moon and stars were so bright that the two animals managed to find enough food to keep themselves alive.

One day, while hunting lemmings, Toolooghak met Teregeneuk. "This darkness wearies me. I have a hard time finding enough to eat," Toolooghak said.

"I cannot agree with you," responded Teregeneuk. "In

this darkness I can slip up on the lemmings and catch them easily, for I am white and the lemmings cannot see me. I like this country."

"Well, for my part, I think I like some sunlight better," said Toolooghak. So he got himself a pair of wings and flew off to the South.

.   .   .   .   .

The story of Toolooghak and Teregeneuk is, you perceive, a short story. But no one in the *tupek* appears to think it comical for its shortness.

Naïve, too, isn't it? Practically no plot at all. You and I, considering ourselves persons of discernment and acumen, are already wondering what subtle significance so short and dignified a tale may have. "Have we here the Eskimo's concept of the beginnings of bird migration?" we are asking. "If so, why select so sedentary a species as the raven? Or is the Eskimo tersely disposing of the whole involved matter of evolution of higher forms from lower forms in this simple decision of the Raven to get himself a pair of wings? Or is the tale an allegory? Perhaps these two beings, the Raven and the Fox, are Ambition and Complacency; the Discontented Visionary and the Vacuous Fatalist; the Explorer and the Stay-at-home."

No one in the *tupek* asks any analytical question. But a bright-eyed boy, in a quiet voice, repeats: "So Toolooghak got himself a pair of wings and flew away to the South." He is memorizing the story.

The *angekok* is speaking again. Now he is telling of Teregeneuk, the Fox, and Amaughuk, the Wolf. Obviously this is to be a thrilling tale, for the children wriggle with delight, and the old persons smile benignly. Everyone in the *tupek* hopes that he is soon to hear, once more, of the victory of the little fox over that most cordially hated of Arctic creatures, the exceedingly treacherous and exceedingly cunning wolf.

. . . . .

One day Teregeneuk, the Fox, was fishing at the edge of the ice. He caught a fine *ichalook*-trout and ate it. Just as he was finishing his meal, Amaughuk, the Wolf, came by.

"How do you do?" said the Wolf. "I see you have caught a very fine trout."

"Yes, I had rather better luck today than usual," said the Fox.

"Would you mind telling me how you catch such fine fish?" asked the Wolf.

"It really isn't difficult at all. You just sit at the edge of the ice, letting your tail dangle in the water. Soon a fish will come and bite the hair at the end of your tail. Then you must leap forward and jerk the fish out before it swims off."

"Thank you," said the Wolf. And the Fox went away.

The Wolf seated himself at the edge of the ice with his tail hanging in the water. Before he knew it, a cake of ice had drifted in, and his tail was caught and frozen firmly. He was very angry. He waited a long time, hoping that the ice would break away and let him go free; but he was held prisoner. Finally, to keep from starving to death, he had to jerk away. His beautiful tail, the tail of which he had been so proud, was gone! That Fox!

Following the tracks of the smaller animal, and becoming increasingly angry and increasingly hungry, the Wolf went inland. After many days of travelling (he was by this time in a towering rage), he met the Fox, sitting quietly in the snow.

"How do you do, Fox? Are you by any chance the Fox whom I met recently at the floe? The Fox I met at the floe told me how to catch a fish."

The Fox nodded his head in greeting and was silent for a moment. Then, reaching out his paw, he plucked a tiny, dry willow leaf that protruded from the snow. This he held up in front of his face.

"I am very sorry; but I will not be able to help you find the Fox of whom you speak. I haven't been able to see anything for several days, for I am snow-blind."

· · · · ·

The story is ended. There is a burst of laughter. Everyone is so pleased that the little Fox has outwitted the great Wolf. The story-teller himself joins in the merriment as if, while telling the story, he had feared some new development that would mean trouble for the Fox.

You, too, probably are smiling. The promptness with which the story ends, the complete adequacy and finality of the Fox's last statement, the extreme restraint of the angry Wolf—these features are comical indeed. The yarn is frightfully inaccurate, of course. Arctic foxes rarely, if ever, catch fish, at least in winter. Nor are Arctic wolves especially interested in fish. The conscientious mammalogist might spend sleepless nights over such misrepresentation of fact.

The point to be remembered is that the Eskimo, disliking the Wolf above all other animals, vents some of his spleen in telling stories wherein the Wolf suffers hunger, ignominy, and defeat at the hands of a much smaller, weaker, and actually less intelligent animal.

Since the Eskimo is himself instinctively polite, it need not surprise us that the characters in his stories are also polite. A weasel-hero may be killing a lemming, but he does it politely. Conversation between the killer and the being-killed is dignified, thoughtful, almost as if nicely chosen words, spoken in a pleasant voice, were a specialized sort of anesthetic.

But we must get back to the *tupek*, for another tale is being told. This time, in quite another tone of voice, the *angekok* is telling of Ookpikjuak, the Great White Owl, and Little Shik Shik, the Ground Squirrel of the Repulse Bay Country.

OOKPIKJUAK AND SHIK SHIK

Once, on a day in late summer, Little Shik Shik was eating some grass not far from his burrow.

Suddenly Ookpikjuak, the Great White Owl, appeared, and before Shik Shik could turn and run, the Owl had caught him.

"Ookpikjuak, please let me go," pleaded Shik Shik.

"Ah, no! You are very fat and I am very hungry," replied the Owl.

"Well, if you must eat me, you must. I am fine and fat, for I have been feeding every day for a long time now, getting ready to go to sleep for the winter. You will have a wonderful meal. Do you not think it would be fitting for you to dance by way of celebration, if I am willing to sing while you dance?"

At first Ookpikjuak was obstinate. "I am very, very hungry," he said. But finally he consented to celebrate with a dance.

So Shik Shik, freed from the great talons, and standing to one side, began singing (what must under the circumstances have been not too joyous) a little song. And Ookpikjuak danced. One stanza, two stanzas, three stanzas Little Shik Shik sang.

Then he thought to himself: I told Ookpikjuak I would sing for him. I have already sung quite a bit. Should I sing any more? Yes, perhaps I'd better sing some more. So he sang another plaintive stanza by way of fulfilling his promise completely.

And great, bulging-eyed Ookpikjuak danced and danced, hopping about on his puffy feet.

Then Shik Shik dived between Ookpikjuak's legs, down into the safety of his burrow.

Ookpikjuak stopped dancing. He was disappointed and angry.

"Come out here, Shik Shik; I want to speak with you a moment," he boomed.

Shik Shik thought about the matter. Then he answered, "No, I think I'd better not come out."

Ookpikjuak waited, wondering what to do. Finally he said, "Come out Shik Shik, your Father is here, and he wishes to speak with you."

This statement worried Shik Shik. If his Father wished to speak with him, he really ought to go out. He thought about the matter for some time. Then he said, "No, I think I'd better not come out."

Then Ookpikjuak thought again, this time for a long, long time. And Shik Shik was considerably distraught. But he knew that he had sung for the Owl even as he had said he would.

At last Ookpikjuak said, "Come out, Shik Shik, your *Great-Great-Grandmother* is here, and she wishes to speak with you."

Ookpikjuak had spoken in such an impressive voice, and the announcement of his Great-Great-Grandmother's arrival on the scene was so overwhelming to Little Shik Shik that he actually started out. Then he paused. A summons from one's Great-Great-Grandmother is not to be disregarded in a flippant manner. He thought about what Ookpikjuak had said for a very long time. He started out again; then he turned back. He scarcely knew what to do.

Finally he said in a faltering voice: "No, I think I'd better not come out."

.   .   .   .   .

A hush has fallen on the *tupek*. A full moment's hush, then hilarious applause. Little Shik Shik has really won! The Great White Ookpikjuak will not get him now! Even the summonses of the Father and the Great-Great-Grandmother didn't work. Everybody is happy because Shik Shik is safe in his burrow.

You will think, as you witness the telling of a story of this sort, that the yarn has never been heard before. Interest in the development of the plot is so deep that everyone appears to be wondering what will happen next. It is almost as if the *angekok* were in the control of some Great Raconteur whose whim or fancy might wreck and upset the whole tale.

The Eskimo is a good naturalist. He is obliged to be a good naturalist if he is to be a good hunter. He has names for all the birds and mammals of his world, and he knows the habits of most of these creatures well. But when he composes his stories he is first and foremost a dramatist, and Natural History simply has to take care of itself.

In the story of Ookpikjuak and Shik Shik, fact and fancy are commingled wholesale. Snowy owls do catch ground squirrels, to be sure. Ground squirrels do eat grass, and become fat in late summer, and hibernate. But beyond these few facts the story is, in the main, a portrayal of Eskimo courtesy, ancestor-worship, and sense of justice.

There is another Aivilik story about Ookpikjuak that I heard several times while I was on Shugliak. This is a story of Ookpikjuak and two Arctic hares.

. . . . .

Ookpikjuak was once lucky enough to come upon and catch two *ookalik*, two hares, at the same time. He caught one in one foot and one in the other foot. The terrified *ookalik* tried desperately to run away. They were so frightened they screamed loudly. But they did not say a word to Ookpikjuak.

Finally they started running off, but Ookpikjuak clung to them, flapping his great wings.

As the two *ookalik* ran, one went to one side, the other to the other side of a great boulder. Ookpikjuak was struck full in the chest and killed.

The two *ookalik* stopped screaming and ran away.

. . . . .

The breath-taking rapidity of this story always amazed me. Every time I heard it I expected to learn a few additional details, but in this I was disappointed. The words apparently had been definitely memorized by most of the men.

After I had listened to the story six or eight times I decided to risk indefinite danger by asking John Ell just what it all meant. The moral might be deduced, I reckoned, that one should not bite off more than one could chew, nor count one's chickens before they were hatched, or something of that sort. But I was interested in the Eskimo aphorism fitting the case.

John was puzzled. He was not used to explaining stories. Finally, he said: "The story means that you should watch where you are going!"

After this I decided it would be better not to ask questions. I would let old Angoti and By-and-By go right on with their tales of fish-eating foxes, fulmar petrels wearing snowglasses, caribou with sealskin boots on their feet, loons paddling canoes, and ground squirrels singing ditties for snowy owls to dance to. I would smile when my friends smiled, laugh when they laughed, become sober when they became sober; and I would do my best to look as if I understood exactly what each story meant.

Too, I would dream of the day when I might tell, in return, such a story of a Raven, a Fox, and a Piece of Cheese as would make them wonder the rest of their lives what it really was all about.

But I haven't told that story yet.

Aivilik hunter with his harpoon.

By and By and Tommy Bluce with Eskimo children.

Doctor Sutton, far right, shown with John Ford and a walrus aboard ship.

Home from Southland: an early Snow Bunting.

Herring Gulls.

Eider nest.

Willow Ptarmigan cock and hen.

Red-throated Loon on nest.

Guillemots.

A chronicle of changing seasons in the mother Willow Ptarmigan's back.

Willow Ptarmigan chick.

A nest of young Snowy Owls with a lemming nearby for food.

Dryas blooming in the wake of the retreating snow.

Aivilik village of igloos and tupeks in the early spring.

An early spring nest with six Snowy Owl eggs.

Chapter XXXVIII

Easter Warfare

NOBODY struts up and down any fashionable avenue on Shug-liak's Easter Sunday. Nobody buys or sends any flowers. But there are the most beautiful snow-white "bunnies" in the world, if you look carefully for them along the ridges; and there is a special service at the Roman Catholic Mission that is attended by most of the Eskimo "converts."

It is my firm belief that these "converts" enjoy the Mission services. And thus enjoying them, they benefit by them. I believe that they like the missionaries, for they recognize sincerity and kindliness when they meet it. Perhaps they like Christianity also, whatever their muddled concept of Christianity may be. I think they must have serious difficulty in understanding anything about such words and phrases as *Magnificat, Gloria in Excelsis,* and *Lamb of God.* The Bible, after all, was not written for Eskimos. Where, in Holy Writ, is there any talk of spirits in the fog, spirits in the aurora borealis, spirits in willow-bushes and gulls and sculpins? These spirits are actualities to the Eskimo. His world is full of such actualities. A religion that does not recognize these can hardly be logical, sensible, thinkable.

The missionaries have no easy task in explaining Christianity, in translating the *ideas* of the Bible into Eskimo ideas. Abstract ideas are often difficult to express at best; and they are especially difficult to express to the Eskimo, whose mind is not used to dealing with such ideas; who cannot, for example, see why any day of the week should be "holier" or "better" than any other day so long as there is no dependable and unvarying abundance of walrus or caribou or seals *on that day*. Or so long as blizzards blow as fiercely *on that day* as on other days. Or so long as dogs die, or children become sick, or persons drown *on that day*.

But I have not set out to discuss here the Eskimo's concept of Christianity. I want to tell you about the preparations for our Easter festival there at the post, and about the row Sam and Jack and I had on the night before Easter Sunday.

Sam sent out word to all the Eskimos, Aivilikmiut and Okomiut alike, everybody on the island in fact, to come to the post for Easter. We were to have a gay time: games, contests, food for everybody, a dance or two at the servants' house.

Sam and Jack and I were busy. There were wagon-loads (I suppose I should say *komatik*-loads) of bread to be baked, hundreds of tins of meat to be carried from the store, and all sorts of boxes to be knocked open. Sam was up to his shoulders in flour and dough, for Mary Ell couldn't possibly bake enough bread and buns by herself. Jack and I ran a canyon through the snow between the store and the house. John Ell was busy hauling out a huge rope, putting up a marker with a red flag out on the harbor ice, and making some new dog-harness. There was a great bustling.

On Friday, April 18, the Eskimos began to arrive. Tooah-tak and his family came, bringing many fox skins, from Cape Low. Tommy Bluce and his family arrived, with a snowy owl for me. Others came. Sam was busy trading all day. I had more fox skulls, lemmings, hares, owls, and ravens than I knew how to manage. The house was crowded. All the tinned meat that was to be eaten was thawing out under and on and round the kitchen stove. We had to go through the provisions selecting prizes. A barrel of tea was to be made.

On Saturday Sam and I got at the final preparations. We decided to make sandwiches out of slabs of meat and thick slices of bread and big raisin buns. We cut bread and opened tins for about five hours. Our thumbs and index fingers were blistered so badly we had to wrap them in gauze. We cut our hands three or four times while opening the cans.

I am not trying to win your sympathy with this recital of our woes. But perhaps you have noticed that Jack did not help with the sandwiches. Nor did he have any blisters on his fingers that Saturday night, or cuts on his hands. Jack, the spang young dude, was all dressed up in his gayest clothes, playing gentleman. Jack was, I guess, in love. Or at least he thought he was in love. And he *was* handsome, I swear he was, in his new white sweater and his red-topped sealskin boots.

Jack was busy cutting a dash with the ladies. He looked at me blankly that day, eyes turned in my direction, but not seeing me at all. His talk was a world or two removed. In fact, as I look back upon that day, I believe Jack *was* in love, bless his bright and shining heart!

That night, after Sam and I had worked ourselves into a state of physical weariness and mental brilliancy, we all had a row. I mean it was a real row. We didn't tear out any fur nor sink any teeth into eye-sockets, but it was close to that. Of course it was Sam's fault, or Jack's fault, or my fault. Something about films that had disappeared, or cigarettes that had been left burning here or there, or carelessness of one sort or other. Hot words were said. Sam would side with me for a while, then with Jack for a while, then against both of us. All of us were in a snip-snap, crick-crack frame of mind, fairly spoiling for battle. It would have done us good

to claw and bite. We sat there in the half-dark, growling and
barking across the room, so angry the whites of our eyes must
have glowed dull red. I don't remember exactly what I said
by way of criticism or threat. I do remember saying that I
wished to go on a long trip so as to see more of the island.
And I remember being told, in no uncertain terms, that it
would be good for everybody if I *did* go on a long trip.

Now you see what sort of person I really was. And just
how popular! But we didn't hit one another; not once.

I felt simply rotten that night. The row had wound up
with Jack's going out into the snow in a sort of daze, saying
that he didn't know where he was going—that he just wanted
to be somewhere else. And I remember that Sam shouted at
him: "Jack, come back here! You hear me? Come back
here!" I certainly did feel rotten.

Next day was Easter Sunday, you see. We had invited the
Fathers from the Mission down to dinner.

When I wakened, I went down and spoke to Sam in my
usual manner. When Sam spoke to me I knew everything
was all right between us. Sam and I were, you see, philoso-
phers.

But with Jack it was different. This young man had been
hurt. He was angry. He would show us just how angry he
really was. It was a very silent morning, with much looking
out of windows.

At about eleven o'clock I decided I'd try to make things
right. Or maybe Jack did the deciding. I don't know just
how it all happened. I think Jack must have looked at me,
for a change. Anyhow I remember saying: "Jack, even if we
are mad, even if we hate each other, what do you say we fool
the priests into thinking we're just as good friends as we ever
were?"

And Jack turned round with the happiest smile you ever

saw, and said: "Sure, Doc! What do you say we just fools all of them?"

And as I remember our dinner party that day it was a very successful affair. There was much gaiety; quite a bit of badinage about Sam's blisters and my blisters and the way we worked so "hard" all winter. And I found myself liking Jack better than I had ever liked him before.

It was a great day, that Easter Sunday. No anthems, no frock-coats, no lilies: just an Arctic calm after an Arctic storm!

CHAPTER XXXIX

Fête du Jour

IF FOR a moment you suppose that old Shugliak has no social life you must plan to attend one of Sam Ford's Easter festivals. You will wear your necktie and your new *komik*, and in the middle of the busy day of races and games and banqueting you will say to yourself: "Well, all this is certainly rather on the gay side!"

The Easter festival of 1930 began with races across the snow: races for the men; races for the women; races in sacks; three-legged races; wheelbarrow races, if you know what I mean; and races for the children. Prizes were given to all the winners.

Then there was jumping for the men, and wrestling, and arm-pulling and kicking. Jack and I, being longer-legged than anybody, could out-jump the island; and the Eskimos appeared to enjoy our leaping and capering.

Then there was a tug-of-war, or rather several tugs-of-war: one between the men and women; one between the girls and boys; and one between the Aiviliks and the Okomiut. Sam told me that one of his reasons for giving the festival was that it would bring the two races together in a pleasant way, helping them to understand and like each other better. As I watched the men and women playing and laughing I sensed little of unfriendliness or exclusiveness anywhere.

A screamingly funny event was the *komik*-race.[1] The men

---

[1] No pun intended.

formed a ring, all facing outward. Then they sat down, still facing outward, and took off their *komik*. Sam and I gathered the boots, all of them, and jumbled them in a big pile inside the ring. At the signal of the revolver the men swung round, each dashing after his own footgear.

It is not easy to describe the furious head-on collisions and wild scramblings that swiftly followed. It was all something like a dog fight, for it was noisy and free-for-all. Men underneath groaned and yelled and crawled out as best they could, gasping for air. In the centre of the mass six or eight pairs of legs stuck straight into the air: the men connected with these legs were looking for their boots, head down, arms down, bodies held up by the mass of shifting, crawling, fighting humanity round them. The hubbub attracted and excited the dogs, who came bounding up to snatch and run off with a boot if given half an opportunity.

The women also had a *komik*-race. And they went at it heartily. I remember dear old Shoo Fly with her weak bones and the sore on her neck. She was not going to be left out of the fun though she was not young any more. She might win a prize, as well as anybody. She dived in, along with the rest, contributing what hoarse screams she could to the chorus of voices. She was a good sort. No wonder she had been the "belle" of Repulse Bay.

The big event of the day was the dog-team race. Out in the harbor, the leader-dogs all with their front feet theoretically on a starting line, forty teams were assembled. It truly was a wonderful spectacle: such a spectacle as I never expect to see again. The teams were not easy to control. There was much fighting and wrangling between dogs who did not know each other. Most of the teams had six or eight dogs. Nearly every good hunter on the island participated.

The revolver barked. There was a chorus of whining and yelping. Whips cracked sharply, almost as sharply as the revolver. What a frenzy of shouting! Commands of all sorts that bewildered the dogs and sent them charging this way and that into one another, across the *komatik,* through other teams' traces! Sledges overturned to right and left. Drivers were dragged hither and yon. Some of the teams became so entangled among themselves or with other teams that they could not go ahead, and the men had to spend an hour unwinding traces. Eventually certain of the teams got safely through the mêlée and headed for the red flag two miles out on the inlet. Muckik's team took the lead and kept it. Once the starting was over the dogs settled down to work and there was less excitement. We all were glad Muckik got the prize (a bag of flour), for he had not been having the best of luck with his trapping.

I tried to take some motion pictures of these games and races. But the camera behaved badly. It had had a hard fall on a *komatik*-trip across the rough ice, and one of the sprockets was broken.

When the games were over the banquet began. The Eskimos piled into the house and scoffed up sandwiches and tea. I have vague recollections of this meal, for I was exceedingly busy passing out food, filling and refilling mugs, and trying to make everybody comfortable on chairs, on tables, or on the floor. Sam and I took the party seriously. We struggled hard to make it "a success." I remember seeing the women stuffing buns into their *kooletah*-hoods, and chewing up food and passing it to their babies' mouths. They all appeared to me a little nonplussed as they sat about taking what we gave them. The men carried their food outside. The dogs were having a feast royal down by the "oil-shed." Of all the company only Sam and I did not eat. We had lost interest in food.

There was a dance that night. Everybody was there. The "kitchen" at the servants' house was packed round the edges. In an open space in the middle such gay blades as Jack and Kayakjuak and Pumyook and Kooshooak and Tommy Bluce danced a sort of square dance with some of the pretty young women. Everybody in the room was hot, but the dancers were wet as rags, and their streaming faces shone with brightly reflected lights. The house trembled violently; the great stove shook; the tables rattled; and the white fox skins that hung from the ceiling silently waved and swung.

There was a little tribal dancing, something on a more primitive order, with a weird rhythm. The men and women, all facing the same way, formed a ring and clomped round the room. The accordion did not play during this dance. The time was pounded out on a table top.

Music was furnished for the square dances by Khagak and his accordion. And by Sam, who played well. Khagak's melodies were not obvious, though the rhythm was regular and strong. I think the tunes were modifications of jigs that he had learned from the crews of the whaling vessels.

The bewildered and solemn-faced children became drowsy as the evening wore on. They fell to sleep here and there and were stacked up on tables and beds wherever there was any room. I doubt if they enjoyed themselves much.

Jack and I played our harmonicas and sang some songs by request. By midnight I was so tired I went to sleep without taking off my clothes. It had been a very full day for me, chiefly because I had struggled so hard to keep up with the conversation of my Eskimo friends.

Tuesday was quiet enough during the daylight hours. But there was another dance that night.

By Wednesday noon the Eskimos all had started back for their encampments at Native Point, at Munnimunnek, and at Cape Low.

CHAPTER XL

The *Sheenah*

THE deeper channels about Shugliak never freeze over. But during almost nine months of the year there is a thick ice-sheet along practically the entire shore-line, a sheet that seals shut the shallower bodies of water such as Coral Inlet, and that occasionally [1] extends across the whole of Frozen Strait to the north, binding the island to the mainland.

The outer edge of this vast ice-sheet is called the *sheenah*.[2]

The *sheenah* has a frigid beauty all its own. Here there is the same thin brilliance of sun and pallor of sky that are the winter tundra's; here the same jade and azure that are the moon-steeped, shadow-struck whiteness of snow. But here rose-colored spires and pillars and minarets of ice move slowly in and out with the tides. Here purple mist-clouds haunt the shifting channels. Here water, forty fathoms deep, glistens black as fluid obsidian at your feet.

The *sheenah* is not as silent, not as apparently lifeless as the tundra. Here there is an almost incessant grinding of ice and lapping of waves; the *chug* of diving walrus and *plash* of frisking seals; the falsetto screaming of gulls; the gabble of *ughik*-ducks; the whistling of knob-nosed king eiders' wings; the thread-thin piping of red-footed *pitseolak*-guillemots.

The *sheenah* is exciting. Everything about a walrus is ex-

---

1 Once in about every seven years.
2 Literally, *edge*, any sort of edge; loosely, *floe*.

citing, from the coughing, grunting, bellowing sounds he makes to the barrels of blood he pumps out over the ice when he is mortally wounded. *Oogjook*-seals are exciting, with their strange way of floating along just beneath the surface and diving with a graceful curving of back and *flup!* of hind-flippers. The ice you stand on is exciting, for underneath that ice, you remember, is Hudson Bay. The very wind is exciting, for has not that wind been known to break vast sections of the floe-edge off, blowing them out to sea?

KINGALIK: MALE KING EIDERS

During the period from April 24 to "spring," we made three extended trips to the floe south of Bear Island. On the first of these we travelled about fifteen miles from the post in reaching open water. On the second we found that the ice-edge had receded landward considerably. On the third we had to journey but a little way beyond Bear Island. Wind, waves, sun, rain and tide battle with the ice-edge constantly, wearing it slowly away as spring advances.

On my first trip to the *sheenah* John Ell and Tommy Bluce were my companions. On our way "down" we stopped long enough on the inlet-ice for Tommy Bluce to get a *netchek* that was sunning itself near its hole. Before starting to crawl up for a close shot, Tommy drew sleeves made of bear-skin on his arms and tied a square of bear-skin to his shoulder. This square he held up in front of him whenever the seal lifted its head. The patient stalking required over half an hour. The Eskimos did not pour any fresh water over the dead animal's nose.[1]

John told me that when a bear kills a seal it crawls up just as Tommy had, walking forward a way, then crouching motionless while the seal's head is raised. I have never been quite convinced that seals are asleep while their heads are down, but the Eskimos firmly believe that they are, and stalk them accordingly.

At noon we reached Muckik's encampment near Bear Island. Here several of the men were stationed at seal-holes, armed with harpoons. Nearly every man had covered his seal-hole with a thin sheet of ice that kept the seals from seeing him. Through this sheet of ice a stiff wire had been stuck a considerable distance down into the hole. The tip of this wire was jarred by the seal as it came up to breathe and the hunter thus was warned that it was time to raise the harpoon. A few of the men had not covered their "holes" with ice, but had placed a feather or tuft of hair on the water. When they saw this tuft trembling they knew a seal was coming up and poised their weapons for the *coup de grâce*.

Muckik told us he had waited three days at his seal-hole without even seeing a *netchek*. He had a case for his "seal-detector," made from long strips of wood he had got at the post. Here he could keep the wire perfectly straight even on his roughest *komatik* journeys.

By the time we got to Eevaloo's encampment it was late

---

[1] This pouring fresh water over a seal's nose is said to be a custom among many tribes of the Innuit. It is a libation performed by way of propitiating the hunting-gods, in the belief that the seal spends its entire life trying to find fresh water.

in the evening. We put up our canvas tent, using the canoe and *komatik* instead of pegs along the sides, and driving a few pegs at the front and back. We were camped on the thick bay-ice, five or six miles from Prairie Point and about the same distance from the *sheenah*.

I had torn my caribouskin trousers during the day. Eeva-loo's wife mended them neatly for me. She was a kindly, gentle-voiced woman.

In the morning we set out early. The snow was deep and very soft, and the dogs had trouble making headway. It was not very cold.

About four miles from the ice-edge (we could see the *sheenah*-cloud [1] by this time), we came upon a crude *igloo* in which had been cached a ton or so of walrus-meat. The dogs were all for stopping here for the night, but we dug out some slabs of meat, lashed these to the *komatik*, and toiled on.

Finally we reached a spot near some rough ice, where we decided to put up the tent. This time pegs were sunk all the way round, for we should need both canoe and *komatik* in our hunting. We had trouble finding snow that was not too salty for tea. Tommy Bluce and I took the team back to the walrus cache and fetched a huge load of dog-food. We fetched also three massive, tusked heads and three big paunches.

We piled the slabs of walrus-meat at one end of our tent. I decided to preserve at least one of the skulls and therefore, as a sort of proprietary gesture, put the largest head at the foot of my sleeping-bag. The Eskimos, upon observing this gesture, each put a walrus head at the foot of his sleeping-bag and distributed the three stomachs, giving me the largest.

We had *tooktoo-quak* and thin slices of walrus-blubber with our tea. Each of us fed from his own "specimen," shav-

---

[1] A gray cloud that gathers over open water at the ice-edge.

ing fat off the cheeks, the jaws, and the upper part of the face with our big knives. We ate a great deal.

I hankered for those raw clams in the stomach near me. Now and then I gave the gray-green mass a jab with my knife. Finally I was probing and poking, separating one clam from another. When I perceived that the molluscs had not been chewed up nor digested to any great extent I tried one. It was delicious in an unexpected way. I tried another. Then I dismissed whatever inhibitions remained and downed them by the dozen. I believe my system needed them, for they seemed to satisfy an indefinite craving. I was amazed that the walrus, in eating all these hundreds of clams, apparently had swallowed only one or two tiny pieces of shell. The Eskimos told me the animals dug the clams up with their tusks, but "shelled" them with their powerful, slanting, molars.

We found, not to our great surprise this time, that we had forgotten snow-shovel, gun-oil, tape-line, iodine, scalpels and several other more or less important items. I may as well forewarn you, if you contemplate spending a year with the Eskimos, that you will always be forgetting things. Forgetting things is an important part of the North Country winter.

We played some "Fi' Hunnun" by way of passing the time; I wrote for an hour or two in my diary-book; then passed round some red candy I had brought.

Next morning we had more *quak* for breakfast, more *quak* and *aiviuk-ookshook*.[1]

It was a windless day. Our *komatik* sped swiftly to the *sheenah*, for it was lightly loaded. We got three *netchek* without trouble. The carcasses floated for they were thickly blanketed with blubber at this season.

I have little idea how far we travelled that day. I fell

---

[1] Walrus-blubber.

through the rotten snow-crust so many times that my shins went fairly numb. I wearied of the gray stretches of "leather-ice" that sank sullenly. We were out ten hours. Sunset turned our whole world an unbelievable golden-orange. Tommy Bluce and I saw a solitary *kakkoodlook* [1] beating its way across the channel just as the sun was going down. The water was red as molten lava.

We had *netchek*-blubber and *tooktoo-quak* for supper. I didn't like this *netchek*-fat nearly so well as walrus-fat. There was no getting away from that *netchek*-odor.

The weather turned bad. For almost a week we hardly stepped outside our tent. On one of these stormy days there was a partial eclipse of the sun. I expected the Eskimos to say something about the doings of the spirits, but they made no quaint comment and appeared to be not in the least disturbed, even when I talked to them about the dim light and the strange shape of the sun.

When the sky cleared and the north wind opened the floe by blowing the loose ice out to sea, we emerged. It was the morning of May 1. Eight of us hunted together, Muckik, Jasper, Billy Boy and two other Eskimos having joined us the night before.

By eleven o'clock we had shot several *netchek* and were happy over our success. We sat on the snow chatting and making plans for the afternoon. Suddenly Muckik hauled out his great knife, ripped open the seal that had most recently been killed, tore out the mass of intestines, and lifted daintily forth the steaming, quivering liver. We ate it. I confess it was not easy swallowing that raw liver, for I did not, as I have tried to tell you, like the odor of a *netchek*. But once I had eaten a little, it was easier to eat more.

About noon I went to sleep sitting in the snow. I had been

[1] Atlantic Fulmar, *Fulmarus glacialis*.

waiting for seals, rifle in hand, on a big, comfortable "chair" at the very edge of the ice. When I awoke the blackness of the water about me fairly made my hair stand on end. What if I had fallen in! Or the chunk of ice broken off and carried me out to sea! Along the floe you are always thinking thoughts of this sort.

I was with Tommy Bluce most of the afternoon. Our dog-team got away from us in the rough ice. We had a hard run. But we finally caught them. And Tommy and I coughed horribly the rest of the day after our unusual exertion.

We returned to the post on May 3. On the way in we visited Muckik's encampment. Muckik's wife was carrying about in her *kooletah*-hood a baby about a year and a half old: a beautiful child with great black eyes. I never saw this baby anything but stark naked. I never saw it cry. It spent its time kicking and bouncing and chewing its thumbs and laughing. Its chest was very flat, however, and it died a few weeks later.

Several of the children at the encampment were snow-blind, some in utter misery, their faces streaming tears, their eyelids red, swollen, and scaly.

The men had got several half-grown *netchek*. These animals had patches of shaggy white baby-fur clinging here and there to the head, back, and sides.

A mile or so from the post John Ell put on one of his wife's red dresses that he happened to have with him. We made quite a sensation as we came whooping into "town."

Chapter XLI

Games at the Floe

It was good to see Sam and Jack again. I was much in need of soap, water, and a hair-cut. Since Sam and Jack also required a barber's attention, we rounded up clippers, scissors, combs, and big Company towels, and set to work. Heaps of hair soon lay on the kitchen floor. This Mary Ell swept up and carried to the servants' house. And there was a fierce new odor about the post for days.

. . . . .

Sam told me that someone had brought in a young walrus for me. Thoughtful as he always was about such matters, he had not let anyone skin the carcass until I had examined, measured and sketched it.

. . . . .

"Well, Sam, do you think Winter has come yet?" I asked, by way of showing how imperturbable I had become concerning the passing of the seasons.

"Doc, it's Spring now—sure enough Spring," Sam answered. "Just yesterday I noticed the partridges [1] are getting brown on their heads."

. . . . .

On May 5, the post was visited by a small company of Netchilik Eskimos that had been living in the Duke of York

[1] Willow Ptarmigan, *Lagopus lagopus*. Rock Ptarmigan become gray, not brown, in summer.

Bay region. Shookalook had told us about these Netchilik-
miut [1] back in early April, and we had been expecting the
visit. In the party were three men, a boy, an old woman, a
young woman, and a girl. They had a good team of dogs.

AHIGIVIK, WILLOW PTARMIGAN, IN WINTER PLUMAGE

The men were the tallest Eskimos I have ever seen, and
one of them had startlingly clear, blue eyes. They spent a
long time in my workroom, watching me. They amazed me
by saying that they had never before seen a baby walrus. One
of them was wearing on his back an amulet made from caribou antler-"velvet": a strange ornament cut out and sewed
in the shape of a woman's *kooletah!*

We learned that these Netchiliks had come across the ice
of Frozen Strait from Melville Peninsula about a year before. The wolves had killed some of their dogs. They had
heard about the trading post long ago, but had not known
how to reach Coral Inlet. They had met our curly-haired
Shookalook quite by chance, and Shookalook had told them
how to reach the head of South Bay.

[1] *Netchek*-hunting people. A distinct tribe of the Innuit living principally to the
north of Southampton Island.

They visited us for two days, then started back for Duke of York Bay. While at the post they lived in their own *igloo,* and ate their own food. They did not ask for anything. Their straightforwardness and independence I admired very much.

· · · · ·

On May 7, John Ell, Tommy Bluce and I returned to the floe. We stopped at Muckik's camp for a short time *en route,* reaching a well-known camping-spot by about five o'clock in the afternoon.

Here we found Jack, Kayakjuak, Jasper, Angalook, Kooshooak, Piahlak, and Pumyook disporting themselves in the snow. It had been impossible, we learned, to reach the *sheenah* because of the roughness of the ice. We put our tent up in no time and had an American baseball game before dark. Our ball was made from old socks and our bat was a plank from the "flooring" of a *komatik;* but we had perfectly good "home" and bases, and it was quite a game.

After supper we played cards. A trifle weary with "Fi' Hunnun," I suggested that we try *Snap.*

Snap is a mild little game with the easiest rules ever. All the players (any number may play) sit round the greasy box-lid that is the table. All the cards, no matter how many decks you have, are dealt. In front of each player there is now a pile of cards, face down. The playing begins. Cards are played face up in the middle of the table, one man flipping his card in, then the man to this player's right flipping his card in, and so on. When two cards of the same denomination come one after the other (two sevens, say; or two Jacks) everybody grabs for the pile of cards that have been played, trying to get his hand on the cards first, and at the same time shouting "Snap!" The object of the game is to get *all* the cards.

Doubtless you have played the game, though you may have called it by another name. It is a simple game. Childish in fact. But just you try it, for one whole evening, with the Eskimos!

All goes well until John Ell or Kayakjuak or Kooshooak becomes really interested. One by one most of the players are weeded out as their piles of cards are exhausted. Now only a few "good" players remain.

Blood, figuratively speaking, is in every eye. Blood, quite literally blood, is on every hand. Finger-nails are split and broken, palms gashed and scraped, wrists jabbed, and every time "Schnap!" is called there is a wild pulling and wrestling and pounding and yelling that sets the dogs outside to fighting.

Wait, please, until you have played Snap! for six hours at a stretch; wait until you have put iodine on the wounds and bandaged up the fingers and hands of six or eight players, before you call an evening of Snap! with the Eskimos a stupid evening. Personally I should call Shugliak Snap! little short of warfare.

On May 8, the thermometer registered twenty above zero. It was almost disagreeably "warm." Clouds covered the sky. We made our way to the *sheenah* through the roughest, most treacherous ice I ever saw. We found no *sheenah*, in fact. A southwest wind had packed South Bay.

We played in the snow while waiting for a change of tide. First it was leap-frog. Then some wrestling matches and races and a snowball fight. Then Horse and Rider, with Jack and me as horses. Then a race that we called a "dizzy" race.

In this race the contestants take sticks of wood about three feet long, stand them in the snow, put their hands on the upper end, then bow their heads over until their foreheads are resting on the backs of their hands. At a given signal they begin running round the stick, remaining in this bowed-over attitude, while someone counts One! Two! Three! and so on up to Ten! At the count of Ten! everyone does his best to beat the others across a straight line that has been marked in the snow not many yards away. If you have not tried this sort of race you will be amazed at your behavior when you start for that line. You charge blindly this way and that. You turn cartwheels. You are lop-sided and off-centre, but you somehow keep going. Whether you are headed for the line or not you scarcely know, but you keep going. You wind up in a heap, burrowing your head and shoulders in the snow. You manage to sit up. Your opponents are lying here and there beside you, or under you, or on you, in the most amusing attitudes, and laughing their heads off. And the gallery is cheering madly.

The Eskimos loved the game. They enjoyed getting dizzy. But be careful, when you play, to select a spot where there are no rocks.

All day we kept seeing huge flocks of migrating *mittek* [1] flying northward. These *mittek* were of two species: the handsome, hump-billed king eider, or *kingalik* [2] (meaning *He has a nose!*); and the so-called northern eider, or *amau-lee*. [3] The flocks appeared to be lost. They swung this way and that, evidently trying to find open water where they might rest and feed. They flew close to us. The Eskimos said they came toward us because they believed our presence indicated an open channel of some sort.

When water-lanes began to open with the going out of the

---

[1] Eider ducks; a general term.
[2] King Eider, *Somateria spectabilis.*
[3] Northern Eider, *Somateria mollissima borealis.*

tide, *mittek* swarmed everywhere. Kooshooak and I went after specimens, paddling about in our canoe. We had no difficulty in getting a dozen or so. The Eskimos astonished me by biting the peach-colored "noses" from the bills of the male birds. These they chewed up and swallowed, smacking their lips. I lost no time in warning everybody that the "noses" were *not to be eaten from my specimens*. This fanaticism was accepted without a murmur. The Eskimos were used to "the Doctor's" strange ways by this time.

In the afternoon there was an eerie booming of thunder, and a sulphur-green flashing of lightning. The Eskimos looked solemnly at one another as the unfamiliar sounds reverberated, echoed from the ice pinnacles, and rumbled across the sky. Obviously the spirits were disturbed, for they were expressing themselves in an unusual way.

In the evening Billy Boy came into camp full of talk about a *tingmük* [1] he had seen in the sky. This *tingmük*, whatever it was, had flown like a great bird. Smoke had trailed behind it. It had made a dull roaring sound. It had been headed east. Billy Boy must have seen an airplane. We might, ourselves, have seen it had we been hunting a few miles to the southeastward. I have never learned what wandering pilot made his way across old Shugliak, on the eighth of May in the year 1930. I wonder if I shall ever know.

The Eskimos shot some walrus during the next few days. Tommy Bluce and I got to the scene of the killing in time to wade about in the thick blood and join in the clam-eating. I saw several walrus, too, crashing up through ice three inches thick, breaking it with their solid heads or with blows from their great tusks.

One day I shot an *akpah*, [2] a Brünnich's Murre, at the ice-

---

[1] Something up in the air; a noun, but not definitely the name of anything.
[2] *Uria lomvia.*

edge. Wind and tide were in my favor and the bird drifted in. As I picked the specimen from the water a section of ice gave way and I had to do some tall scrambling to keep from falling in. Managing yourself on an ice-pan makes you think of an elephant balancing itself on a rolling barrel at the circus.

I shot an *oogjook*-seal with my Krag. The minute the creature began bleeding John Ell's team came on the run. They knew there was a meal in the offing.

Not everything that "the Doctor" ate these days agreed with him. We had had a meal of raw *netchek*-liver one morning and the Eskimos decided it would be fitting to celebrate with a "dizzy" race. Of course "the Doctor" had to join in too, for being much taller than anyone else "the Doctor" was exceedingly funny as he cut his dizzy capers. "The Doctor" raced and won, as he frequently did. But he also became sick and had to forgo the pleasure of retaining his *netchek*-liver. The dogs, hungry as they always were, didn't in the least object to the second-handedness of a meal. They stood there at "the Doctor's" bedside, holding his hand, a look of genuine concern on their wrinkled faces, waiting for whatsoever the Gods of the Disappointed Esophagus had to offer.

Chapter XLII

Itiujuak

In mid-May old By-and-By had a strange dream that made him think he was going to die. I tried to learn something about this dream, but all By-and-By could say was that he was sure he was going to die.

Some of the Aivilik men talked at length with Sam about the matter. Sam did not laugh nor smile. They all decided it would be a good idea to take a new carpenter's brace and bit down to By-and-By's *tupek* at Bear Island. This new brace and bit were to ward off death.

Was By-and-By's *tupek* the scene of weird religious rites? Were long prayers offered to the great Sedla, Goddess of the Sea? I do not know. I probably shall never know. At any rate, By-and-By did not die.

.   .   .   .   .

On May 19, Tommy Bluce and I set out with *komatik*, nine dogs, a *netchek* for dog-food, a tent, and a week's rations for two men, bound for Fox Channel. I wanted to visit the northeastern shore of the island: the cliffs that breasted the sea; the Porsild Mountains; Canyon River, if possible.

We started about ten o'clock in the morning, heading for the cliffs at Itiujuak,[1] eighteen or twenty miles east of the post. The wind was steadily rising. By mid-afternoon the

---

[1] *Itiujuak*, as a common noun, means *big cliff*. At Shugliak the word is now a proper noun referring to the high cliffs east of the post.

snow was drifting badly: typical Shugliak weather, but some-how so strange for May!

In the middle of the tundra the dogs suddenly nosed a fresh fox-trail. Without a word from us they dashed off, turning this way and that for a time, then settling down to a straight course. Within fifteen minutes they came to a stop. The leader-dog sniffed at the snow. We got off the sledge and ran forward. There, in a cozy burrow, was a fox, its golden-brown eyes glowing. Tommy Bluce caught the animal without difficulty, killed it by kneeling on its ribs, and skinned it out as we continued our journey.

The animal was not in perfect winter pelage. Already the backs of its ears were turning brown, and the inside of the skin, the "vellum," was blotched with dark-gray: sure signs of Spring.

When we reached Itiujuak we found much of the ground at the base of the two-hundred-foot cliffs bare of snow. Here a flock of about a thousand snow buntings were feeding. The pretty birds, nearly all of them brightly marked males, fed in a sort of revolving mass, the individuals at the rear flying constantly to the head of the flock and settling just in advance of the foremost row. They headed into the wind, for turning too much to one side or the other meant being blown upside down.

A hardy duck hawk frequented the cliffs, apparently living upon the buntings. I shot this hawk.

We erected our canvas *tupek* on a smooth drift in the shelter of the cliff. In the course of establishing ourselves for the night we discovered that we had forgotten snow-knives, screw-driver, file, and all the tent-pegs. We could not make an *igloo* without a *pana*-knife. We should have to anchor our tent with boulders. We all but broke our feet kicking boulders loose from the snow.

To my dismay I found that my shot-gun would not "break." I finally took it apart, using pocket-knife and hatchet as a screw-driver. When I put it together again only one of the triggers would pull.

While climbing a hill that evening I slipped on the icy crust of a long drift and wound up in a bewildered heap at the bottom. I should have to be a little more careful in this hill-country. There were high cliffs here, with twenty-foot drifts at the bottom in which a man might suffocate.

We had had a full day and were glad to get to bed.

In the morning we broke camp early and headed northeast. But once we reached high land we found ourselves facing a terrific wind. The dogs had trouble keeping their feet. They could not go ahead. We turned back. We picked the most sheltered place we could find and put up the tent again.

The gale that raged the next three days was the sort of gale you remember. Our tent was in a sheltered spot, but the wind swooped down and about the cliffs, flapping and bellying the canvas and pushing the frail thing farther and farther forward. The ropes, which were very old, began fraying and snapping. We mended them with wire I had brought for preparing hare specimens. Snow piled along the north and west sides, bulging the walls in and threatening to disassemble the entire structure.

The buntings sought windless niches in the cliff-face. We could see them perched cozily here and there high among the rocks.

My most vivid recollection of the storm is of the snow-sheet that silently, endlessly curled over the crest of the cliff far above us: a noiseless, graceful, but somehow sinister cloud, like a steam-cloud from some high hot spring.

It was not very cold. In spite of the wind we tried a little hunting. Tommy Bluce traversed some of the lower ridges and got a hare. I followed the base of the cliff, shooting some small birds and several ptarmigan.

I came upon the ptarmigan, as usual, unexpectedly. I chanced to see the form of one bird clearly chiselled against the darkness of a boulder; then, looking about me noted dozens of the pretty creatures running this way and that. I had been walking amongst them for a long time. They were all round me. When I shot, the flock whirred noisily to the crest of the cliff. Here, meeting the terrific blast, they rocketed upward and off into the opaque, snow-spotted grayness that was the sky.

Our tent held together somehow. Tommy Bluce replaced some of the ropes with dog-traces. I wrote a good deal, using a pencil, for I couldn't keep the ink thawed. I sang Annie Laurie and In the Sweet By and By for Tommy Bluce, and he sang some Eskimo songs for me, among them Khagak's *Ai Yi Yi!* [1]

Tommy and I had a good time together. I showed him exactly how I wanted him to skin bird specimens when he should go to Cape Kendall after blue geese in June. And he showed me how to carve *tooghak*-ivory. Tommy knew no English. But we got along. He told me a story or two and described a *kayuk's* [2] nest so carefully that I knew perfectly well what sort of bird he was talking about.

I tried telling him an Æsop's Fable or two in Eskimo. And Tommy, being a gentleman, laughed as if he knew what I was talking about.

I had a piece of old magazine with me. On one of the

[1] *Ai Yi Yi!* was Khagak's favorite song. I think it was one he had composed himself. It was an endless repetition of the phrase *Ovoongah, Ai Yi Yi!* meaning roughly *I am happy!* The Eskimos, in singing this, took a deep breath, then sang as rapidly as possible, seeing how many times they could repeat the *Ovoongah, Ai Yi Yi!* before they had to take another breath.

[2] Rough-legged Hawk, *Archibuteo lagopus sancti-johannis.*

pages I came, quite by chance, upon a bit of doggerel by Thomas Godley (I think this was the name), a sort of burlesque upon mediæval Latin. It ran:

> "What is this that roareth thus;
> Can it be the motor-bus?
> Yes, the noise and hideous hum
> *Indicat motorem bum!*"

I read this to Tommy and he laughed uproariously. I am certain he did not understand a single word. But it did have a funny sound and Tommy was not going to fail me when he perceived that I was trying to help him have a good time.

On May 22, we decided to strike back for the post. The wind was savage, but at least it was not a head-wind. We fed the dogs the last of the *netchek* and started. It was not an easy nor a pleasant trip. On the way we saw a snowy owl crouched behind a boulder seeking shelter from the wind.

As we came closer to the post Tommy continued to say to the dogs a quaint little expression. I do not know the exact words of this expression, but it sounded like *Ah! la iglooapik!* Tommy repeated the phrase in his softest, most affectionate voice. He was promising the dogs a "little igloo, all nice and warm," so to speak, if only they would hurry a little more. Tommy also used a favorite teamster's expression that might be written down *"Eenoogliowgookook!"* Since the Innuit have no alphabet and no spelling I have no idea whether I have written this word (or these words) down correctly. I do not know what the expression means. I could never pronounce it correctly. By way of being polite to me, my Aivilik friends sometimes actually mispronounced it just as I mispronounced it. You say it to the dogs when you are angry at them. Perhaps it is profane language.

Tommy Bluce and I reached the post about three o'clock in the afternoon. We were glad to be home.

Tommy Bluce said to Jack: "The Doctor and I had a good time. We did not become lonely. The Doctor talked to me and I talked to the Doctor."

CHAPTER XLIII

Sanctuary

I AM thankful to this day that Tommy Bluce and I struck out for the post when we did. Our poor old tent could not have stood that storm much longer. The gale abated for a time on the day we journeyed in, but it regained its force that night and on the following day was worse than ever.

Windows continued to blow in at the house and store. The powder magazine, the widow Kuklik's shack, and all the little outhouses were practically buried. We could see only about thirty feet in any direction through the storm.

A large flock of snow buntings lingered in the narrow, gravel-lined chasm that the wind gouged out immediately round the house. They had a wretched time of it, what with the voracious dogs, the predatory children, and the violent wind that twisted their thin bodies this way and that no matter where they stood, or shot them high into the air whenever they spread their wings.

Jack and I put some grape-nuts out in the gravel. The buntings quickly gathered, working their way up-wind, trying to keep back of little boulders or chunks of snow.

The dogs continued to chase and catch them. Some of them died of exposure and starvation. The children kept shooting at them from the doors and windows and throwing pebbles at them. The only windless place they could find was the area under the back steps where the dogs were accustomed to curling up and napping.

Jack and I decided we should have to bring the birds inside the house. We got a dishpan, a little stick with a long string attached, a gunny sack, and a butterfly net. We propped the dishpan up on the stick, covered it with a gunny sack weighted with stones on the windward side to keep it from blowing away, put some grape-nuts under it, and came inside. Within ten minutes we were pulling the string, "dropping" the dishpan, and catching our buntings. We sometimes caught as many as five at one time. Finally we had caught the entire flock, perhaps two hundred birds.

Inside the house the birds behaved with surprising decorum. There was no banging on windows, no panic-stricken dashing this way and that. They flew when we came near them, or walked about calmly, pecking here and there, and looking out the windows at the storm. They avoided the top of the hot stove, but did not hesitate to gather in little companies under that big, warm, black "rock." They fed upon bread, grape-nuts, and cornmeal that we set out for them.

Frequently they made their way into my workroom. Here they ran about over the table, perched in rows on the window-sill, or sat on owl and raven specimens, turning their heads comically as they eyed the unfamiliar objects under their feet. One of them succeeded in upsetting an ink bottle. But this worked no havoc, for the ink was frozen. Another looked a long time at a gyrfalcon specimen and finally decided to give a danger signal. The rest of the birds took up the excited chirping for a time, but finally came to the conclusion that the alarm was false.

The buntings stayed with us for the duration of the gale. When the wind died and the sun came out the males burst into song. Even now they did not dash themselves at the windows. As a matter of fact, the windows were so dirty and snow-covered they were scarcely transparent.

When we let our pretty charges go most of them hopped out and flew off with a merry chirping. One of them, with an injured wing, flew straight into the mouth of a waiting dog and was swallowed alive. And one, which I held in my hand a moment, suddenly gasped; there was a choking and rattling of blood in the throat; and the little creature died in my outstretched palm.

Chapter XLIV

Winter's End

THE bright flag of Great Britain flew for Queen Victoria on May 25. It flew for me, also, for I found a snowy owl's nest that day, the first snowy owl's nest I had ever seen.

The six round white eggs were lying in about an inch of icy water in the very heart of a snow-drift ten inches deep. The mother *ookpikjuak* had sat on her nest throughout that long, wild gale.

We had a pale omelet for supper.

The following morning John Ell and I returned to the *sheenah*. At an encampment we visited *en route* we learned that Muckik's wife's baby [1] was very sick, that the Aiviliks had given up all hope of its living. We were relieved to learn that many *netchek* had been killed, so there was no food-shortage.

At the *sheenah* we found a great deal of bird-migration going on: wedge-shaped flocks of Canada geese, most of them of the small subspecies ornithologists now call Hutchins's geese;[2] pairs and small flocks of whistling swans; red-throated *kokshowk;* noisy squadrons of *ughik*-ducks—all steadily passing northward.

On the next day we shot several *netchek* and one *oogjook*-seal. Most of the wounded *oogjook* were sinking now, for their blubber-layer was becoming thin. We ate boiled *oog-*

---

[1] Not Muckik's baby; the child had been adopted.
[2] *Branta canadensis hutchinsi.*

jook's intestine at several meals. In preparing this dish John cut the intestine into three pieces of equal length, put these pieces between the fingers of his left hand, and pulled them rapidly through with his right hand, thus squeezing the partly digested material out. Then he braided the limp, flattened strips deftly, and coiled the braid into a pot. The intestines shortened and thickened and became brittle with boiling, so that when we cut them the sections snapped and popped and sprang this way and that.

I got all the bird specimens I could manage and more; guillemots, some of them already in almost full breeding plumage; eider ducks of two species, and handsome oldsquaws in their bizarre nuptial dress.

We ate fried old-squaw one evening, much to our regret. I noticed shortly after our meal that I did not feel exactly well, but said nothing and went to bed. At midnight I wakened realizing that I was sick indeed. I noticed that John also was stirring.

"Are you sick, John?" I asked.

"Not a-now; but mebbe shoon!" John answered, and forthwith became so in no equivocal manner. We were sick all that night, all the next day, and all the night and day after that. We could take neither food nor drink.

John said laconically: *"Ughik,* she not good!" [1]

We took down our tent listlessly, hoping the dogs would somehow pull us back to the post. Fortunately the weather was good. On the way back I could hear the trumpeting of transient cranes, high in air. We encountered so many *netchek* in one of the crack-lined coves that John decided he would

---

[1] For a time we thought these *ughik* had been eating certain foods that made their flesh poisonous to human beings. We finally decided, however, that arsenic from my taxidermic outfit must somehow have got on the bodies during the skinning process. During our illness neither John nor I could swallow either food or water without vomiting violently.

try to get one. He wabbled off with his rifle and bearskin "blind," while I stayed with the dogs.

The farther away John got the more interested the dogs became. They quieted down when I spoke to them, but their ears continued to lift and their noses to twitch. And they were bound to sit on their haunches instead of lying down. They had not had a hard day and were in hunting mood. Furthermore, they had not eaten any *ughik*-duck. When John's rifle cracked, they were on their feet. I shouted for them to lie down but probably made too much noise in this command and only excited them the more. I jumped hard with both feet on the traces just in front of the *komatik*. But I was too late, or I was not heavy enough, for in an instant I had been thrown off my feet, bounced on the taut traces, and caught under the front of the speeding sledge. I kept on shouting to the dogs. My body, held as it was, between the traces and runners, was an efficient brake. The team stopped. I pulled myself out as best I could. And this time, when the dogs started for John, I muttered an inconsequential oath and let them tear.

At the post, John and I recovered slowly. Within a day or two we were eating and drinking again.

. . . . .

It was now June. The lakes were thawing. Ptarmigan, foxes, hares, weasels: all these were turning gray and brown again. A fly buzzed about. Little black spiders walked meekly on the snow. Every willow bud that the ptarmigan and hares had spared was swollen, and every twig was freshening to yellow-green. But the tundra was buried in snow. And the inlet was a vast sheet of ice three or four feet thick.

Winter died in a four-day convulsion of storm. Once more our world was transformed by wind and snow. Once more

deep drifts formed. Only the chirping of the friendly buntings kept us thinking sun and flowers and skies of June.

．　　．　　．　　．　　．

We wakened on the morning of June 7 to find it Spring.
The damp drifts were sinking away before our very eyes.
Impatient loons were circling the lakes, attempting to alight
on the water-covered ice and yelling like wild Indians.
Myriads of lemmings, driven from their burrows by floods,
were clustered on the rocks, waiting to be snatched up by
the gulls, the jaegers, and the owls that were fattening themselves upon them.

Jack and I had occasion, that day, to eat our luncheon near
a little rivulet that had cut its way through the snow. We
noticed that the bodies of dead lemmings were drifting past
us on the stream. We thought it might be interesting to
gather these. We picked the little carcasses from the water
as they floated by. Within about fifteen minutes we had ac-
cumulated a pile of thirty-two drowned lemmings!

Spring at last! *Oopungakshuk,* the time when all the birds
come back from the South. So sudden was the leap of this
new season upon us that we scarcely had time to comment
on the Winter's passing.

CHAPTER XLV

A Day in June

I THINK I shall let you read two paragraphs from my diary-book. They are safe paragraphs:

"Wednesday, June 18: Too tired to write much. Got up at 3:30 A.M. and went north to the brook and got a black-throated [1] loon, male, and shot a 'Greenland' eider, but didn't get it. Saw one Sabine's gull.

"At about 10 A.M. we left for Itiuachuk to make camp for a while. I walked out and found a goose nest right away. On the way we had trouble—*komatik* swam through water. Dogs got loose and we had to chase them a long way. Got dragged through snow and water on my belly. Saw a good many things."

I wish you might see those two paragraphs as they are in the diary-book. A glance at the all but illegible scrawl would convince you that their author had, indeed, been "too tired to write much."

Provided you have slept soundly for a few hours, there is nothing very difficult about rising at 3:30 o'clock on a June morning in the Arctic. The tundra is steeped in brilliant, if somewhat thin and chilly light. The sky is blue and cloudless, and you cannot help feeling, as you look upward, that the air is very pure and very cold and that the sun may be, indeed, quite ninety million miles away.

The world is noisy. Everywhere about you is the sound of

[1] A Pacific, not a Black-throated, Loon. The Pacific Loon is the American sub-species of the European Black-throated Diver.

water: the roar of turbulent rivers coursing down from the highlands, flooding the moss-covered meadows; the rush of brooks leaping out from sodden drifts, dashing down through the ridges, debouching over the rotten ice-crust of the bay, and cutting chasms everywhere; the murmur of the incoming tide, lifting and jostling the ice-chunks and subduing the eagerness of streams; the susurration of rivulets running down across the quartzite and feldspar crystals of granitic boulders; the *plop, plop, plash, plash* of drops falling from overhanging rocks, from the edges of old snow-banks, and from pinnacles of ice.

Birds are mating. Geese fly about *cronk*ing noisily. Ptarmigan cackle from the ridges. Loons wail and groan and laugh their mirthless laughter from the thawing lake-margins. Gulls scream. Terns fly about high in air, calling harshly, the sun shining through their thin wings. Sober-faced jaegers dash about, crying *error, error,* as if their day had gone all wrong. Hundreds of male Lapland longspurs join in a chorus of flight-songs, leaping up from their low perches and drifting back to the tundra on wide-set wings. Snowy owls hoot interminably, the boom of their hollow voices sounding even from across the frozen harbor, seven miles away. And above all these the horned lark sings, breasting the high breeze or fluttering about in gentle circles, matching the chary warmth of the sunlight and transparent turquoise of the sky with the tinkling far-awayness of his lay.

Everywhere on the ridges purple saxifrage blossoms are opening. Most of the snow is gone, but there is the whiteness, now, of the carpet of eight-petalled *Dryas.*[1] Here and there in the more sheltered places gleam spots of yellow: buttercups and poppies, high on their thin stems. A bumblebee zooms about, pawing the flowers clumsily, getting pollen dust

[1] *Dryas integrifolia,* probably.

on his steel-purple wings. There is no flash of butterfly, nor whine of mosquito—for summer has not yet come.

This is Shugliak's tundra in Spring: this the world I saw and heard at 3:30 o'clock on the morning of June 18, 1930.

I put on plenty of clothing, for there was a breeze from the northwest and fog might blow in. I took my shot-gun, heading "north to the brook."

It was a two-mile walk across the salt-water ice. At the very edge of the inlet I had to leap from ice-islet to ice-islet, for a long pool had gathered from the swollen streams, and the tide was in. I crossed deep, mud-lined gorges the freshets had cut, leaping where I could, and crawling down and over and up where I had to. I threaded my way among tidal pools. Here and there I noted pieces of brown kelp, abandoned "seal-holes," and faint signs of dog-sledge trails. The ice was melting rapidly everywhere; it was dirty and porous; the glare was terrific.

At length I reached the brook. At its mouth the savage stream had gouged out first a fan-shaped basin, then a tortuous ice-chasm leading far out into the inlet. The water was not muddy, but it was heavy with sparkling silt. Above the roar of the torrent could be heard the occasional *thump chug* of boulders rolling downstream, and the hissing and crunching of pieces of ice striking against the walls of the chasm. I did not linger long at the brink. Sections of the wall were giving way with a sullen sound and low splash, disappearing for a moment, then rising to be swung and rolled and eddied slowly away.

Progress was not easy here, for the ice was rough. At the bottom of a shadowy crevasse black water churned up in a direction opposite to that of the main stream. The unrelenting roar gave me a vague feeling of uneasiness. How could I hope to hear the little sounds that warn of danger? Might not the ice of the whole inlet give way all at once, bearing me out to sea?

A flock of Arctic terns were fishing in the basin. With red beaks pointed downward they beat back and forth, cackling harshly at one another, setting their long wings for a sudden dive, plunging straight into the water, then making off with tiny sticklebacks that they frequently swallowed in mid-air. Now and then they all came to rest, facing the wind on a high ice-knob, or lining a low margin of the chasm with the pearl-gray of their mantles and the glossy black of their crowns.

A guillemot flew in from the distant sea, spreading its red feet comically just before dropping into the water. As I watched it I became suddenly aware of a gaunt, snaky form in a quiet eddy not far from me. I turned to look the more closely, for I knew I had caught the flash of a bright eye. There, not fifteen feet away, but headed now for the dim depths of his hunting-ground, was a Pacific loon, another of the feathered fisherfolk. Having seen me from his nether-world he had risen cautiously, not even revealing the full length of his velvety neck nor his broad, checkered back. But I had glimpsed his carmine eye, and the narrow white fluting of his throat. Handsome diver! Sad-voiced *Kudloolik* of the Eskimo stories!

Scarcely had I begun to wonder how I might hope to collect so fine a specimen when another loon popped up near me, transfixed me with a wild and brilliant stare, and made off rapidly, turning its head from side to side. It had been fishing under the ice. Since I did not move, it scarcely knew what to make of me, but it was not going to let its curiosity get the upper hand of its innate timidity.

I sprawled at the edge of the chasm, gun in hand. For the moment I had forgotten the roar of the swollen brook and the rottenness of the ice, for I was hunting. I don't know how long I waited, but those loons must have stayed under fully a minute. When they came up, some distance downstream, I shot the nearer one.

The bird was dead. I could see the big feet waving slowly as the silvery breast turned skyward and the slim body began drifting away. I was up in an instant, found myself racing along the chasm's rim, leaping across cracks, sliding down wet slopes, following secondary gorges until I came to narrows I could span, all the while keeping half an eye on my loon. We were both approaching the end of the chasm. The walls were becoming lower. Here the current, retarded by the mass of ice-chunks caught between the converging banks, was not so fierce. Now I was leaping from ice-cake to ice-cake, thrilled and frightened at this slow sinking, this unfamiliar and shifting slipperiness, this queer sound of hissing and grinding and sucking, this impossibility of knowing how thick or solid or dependable an ice-chunk may be. I didn't fall in. I went through a crust once, but struck another firmer layer. One chunk broke under me, but the adjacent chunk was bigger and I kept going somehow. I got one foot wet and scraped my shins. The chunks turned over maliciously; they broke up after me; they undertook all manner of foul play. But they were too slow about it.

I snatched the bird from a little crevice in which it had lodged. When I reached firm ice again I was glad to rest. It was a beautiful specimen.

But when I started back to the post I headed for the middle of the inlet. It would be a long way round. Going somewhere by way of floating ice-chunks is good fun if you don't think too much about it and if you have a Pacific loon to retrieve; but you don't "wend your way homeward" in any such manner if you can avoid it.

I don't know what time it was when I got back to the post, but I do recall that I was ravenous for breakfast; that I was eager to get at a water-color drawing of my precious loon before his bill and eyes and feet should lose the delicate coloration of life; that I was determined to lose no time in starting on our proposed dog-sledge trip to Itiuachuk before the ice of the inlet should break up; and that I was momentarily bewildered by the appearance at this inopportune time of an Eskimo from Koodlootook River with two great Canada geese under his arms.

Breakfast found its way down all right. A drawing of the loon got dashed off somehow and the loon himself skinned out. I measured and weighed the geese while Jack and Santiana gathered together flour and tea and sugar and cheese and bacon and cornmeal and arsenic and cotton and sleeping-bags and guns and tent and ammunition for the trip to Itiuachuk. The *komatik* was loaded by nine-thirty o'clock. We were off by ten, with much shouting of good-bye and whining of dogs.

Itiuachuk was a gravel plateau about fifteen miles southeast of the post. The plateau itself held no special ornithological interest for me, but the Eskimos assured me that geese of many sorts, cranes and swans nested on the prairies at its base, and that blue geese might also be found.

We were a happy three. Short-legged Santiana sat in front, making comments to the nine dogs, who needed no syllable by way of encouragement. Just back of the Eskimo sat Jack, walrus-hide whip in his hand. I was perched comfortably in the rear, a basin of cornmeal at one side, a pocket-knife in my right hand, skinning out geese. An unskinned goose dangled from the *komatik's* handle-bar back of me.

For about three miles the going was pleasant. The dogs knew enough to keep on the smooth ice, there weren't many chasms and cracks to cross, we didn't encounter any deep tidal pools, and I finally finished skinning and salting my geese. The cornmeal spilled once or twice and I jabbed my thumb with a broken goose-bone; but otherwise all went well. The red roofs of the trading post were not fading in the distance, for the June atmosphere was very clear; but they were becoming small. Now we were approaching an area of lovely blueness on the ice, a section of the inlet that had been flooded by a swollen river. By this time the dogs were so accustomed to getting their legs wet that they forged straight ahead, sometimes losing their balance comically, whining in dismay, and turning about for a word of sympathy; sometimes, where the water was shallow, pulling the sledge along at a merry clip only suddenly to find themselves afloat with the *komatik* charging fiercely into their midst and throwing sheets of water out to either side.

The blueness was becoming more intense. Blueness stretched all about us now. The dogs were beginning to be dubious. Santiana did not smile and Jack's face had an expression of concern. All at once there was no sound of the scraping of runners on the ice, no splash of galloping dogs. They were swimming. The *komatik* was afloat. Santiana, being first of all a normal Eskimo, frankly did not like all this water, and curled his feet up under him. Jack and I leaped off, grabbed the handles at the rear and tried to help the dogs as best we could. They were not whining now; they were paddling for their lives.

The water was frightfully cold, of course; but somehow we didn't notice that. I do remember being glad the sun was shining so brightly. We were in some sort of current now. Would the *komatik* turn over? Should we get aboard to steady it? If we got into water beyond our depth what should I do first, try to save Santiana? Jack said the dogs could take care of themselves so long as the sledge was not pulled under the ice by the current. Cheerful thought!

The going was bad. The ice-bottom of this vast blue pool was exceedingly uneven. Had we not been able to steady ourselves with the *komatik* handle we could never have kept our feet. Santiana, bless his honest heart, was shedding tears. He was only a boy anyway.

"Watch out, Doc! The river, the river!" I heard Jack shout, and saw an ugly ribbon of black cutting through the blue just ahead of us. "Hang on! Don't lose hold!"

I grasped the *komatik* afresh, let my weight rest upon it as the black ribbon glowered under us, felt the sledge sinking down and for an instant had a wild impulse to let go and swim as I felt the main current catch us and sweep us along. For a time I thought we were lost. Santiana was clutching the dog traces fiercely. The dogs were doing their best.

All at once, with a blessed sense of relief that I could not explain, I perceived that we were in the blue again. We had not exactly crossed the river, but the current had washed us downstream onto the ice. Soon we were actually out of the water. Behind us the narrow black ribbon appeared harmless enough. The dogs shook themselves and whined. Santiana wiped his face with his hands. We were a wet and sorry expedition. Even the geese were wet.

We wrung ourselves out as best we could. Jack hauled out and started the Primus stove. We had a little trouble in finding fresh water for tea, but we soon had a hot drink. Somewhere we found some drier, if not entirely dry, clothing, and resumed our journey. I was getting sleepy.

Never was there more glorious sunlight. The wind died down. The air became almost languid. Parts of us dried out thoroughly. The smile came back to Santiana's face. Conversation about swans and cranes and geese was resumed. A flock of brant flew by, gabbling loudly. Evidently they were migrating even at this late date, for they kept straight on, heading for the highland at Itiujuak.

There came a lull in our talk. Perhaps we were dozing there on the comfortable *komatik*. The ice was smooth and the dogs were pulling faithfully. We were progressing slowly, but we were progressing nevertheless. Itiuachuk was plainly a good deal closer.

All at once the *komatik* stopped. Someone shouted a threat to the dogs, but we stayed stopped. I looked up, a trifle bewildered, and was astonished at seeing the team running blithely ahead of us, all of their traces properly bound together, but the whole lot of them, traces and all, broken free of the sledge.

"The dogs!" I shouted. And Santiana came to. Jack began shouting all sorts of Eskimo words and phrases. Some of the dogs turned. Some stopped. Some sat on their haunches. Then these dogs rose and other dogs sat. There was no single idea pervading the team-mind. Eventually they all came to some manner of agreement and trotted off.

I was considerably exasperated. I felt that had I run forward when I first perceived what was happening I might have prevented this escape. But I was so convinced that the dogs would obey Santiana's slightest word of command that it never occurred to me I should attempt any gesture whatsoever. I marvelled that Jack had not done anything. What was going on here? Was this all a farce? Were the dogs circling off just for fun, getting ready to come back and set to work after a little breathing spell?

The dogs had not the faintest notion of coming back. Their spirits gradually warming to the possibilities of the situation, they were galloping merrily now, looking back with broad smiles on their faces. Confounded brutes!

"We'll have to go after them, Doc," Jack said simply, just as if I were not already worn out with this eventful day. I continued to be exasperated.

"Why didn't Santiana stop them? What should I have done? What do you say to the Huskies when they start running off that way?" I asked. And Jack answered by telling me a little tale of a *komatik* trip of this sort that was delayed seven days while a runaway team were caught and brought back for service.

I was longer-legged by far than either Jack or Santiana. Wondering what under heaven I should say or do to the dogs if ever I should overtake them, I toiled on, pulling up my damp, wabbly, flappy boot-legs, throwing off my outside dickey-coat, giving vent through needlessly *sotto voce* expressions to my mounting exasperation, and wondering whether we should ever reach Itiuachuk. Far behind me I could hear faint Eskimo expressions asking the dogs please to stop, begging them to stop, promising rest, cozy little snow-houses and heaps of seal-blubber. "Pointless cajolery" was about what I thought, though the word *cajolery* assuredly never occurred to me. "Why didn't they stop those dogs when they had the chance?"

I looked back at the *komatik*. It was a mere speck now in the distance. Not far from it was the short black form of Santiana laboring slowly forward. Nearer me was Jack. I was a long way ahead. Jack and Santiana continued to make promises to the dogs.

My guess is that the dogs ran fully two miles—perhaps more—before they were foolish enough to run to both sides of a pinnacle of ice and get their tangled traces so caught as not to be able to run ahead. Had they reasoned the matter out, had the proper dogs retraced their steps, had they pulled hard enough or chewed their traces a little, they might be as free as birds to this day. But there were limitations to their thought processes or they didn't care, so they ran no farther. When finally I caught up with them they had had a pleasant rest. All of them were glad to see me.

And what could I say to them? or do? My Eskimo vocabulary was pitiably limited. My best English would be utterly wasted. There wasn't a man-sized club anywhere nearer than the trading post. Every stone suitable for pounding dogs to a pulp was miles away hidden under snow and ice. I kicked the dog nearest me, just because I couldn't resist the impulse. He howled and the rest of the team looked uncomfortable for a fleeting moment; but that kick hurt my already tortured toes so badly that it took real guts to keep from moaning and holding my foot in my hands. Queer how self-conscious one may be out in such a desolate place!

Within the next few moments there was extended to me perhaps the subtlest compliment I ever received in the North Country. When I looked in Jack's direction I perceived that he was returning to the *komatik*. Trusting soul that he was, he felt that I could manage those nine dogs alone. Perhaps he was right. At any rate, in my fashion, I did manage them.

I could see that the dogs all knew they should go back to the sledge. They were restless to be off. I grabbed up the bundle of traces, pulled them loose from the ice pinnacle, said a few choice Eskimo words such as those for "pocket-knife," "snow-knife," and "baby," just by way of showing the brutes that I knew what I was about, and we were off. At first the team did not want to run. We idled along about as a Fifth Avenue lady and her leashed Dobermann Pinscher idle along, traces all sportily taut, dogs all pleasingly vivacious, I with a slightly come-what-may expression on my face.

This peace was to be with us for about five minutes. I heard that whine of eagerness that portends a frisking off.

But what could I do? I spoke in the placidest voice I could muster, bidding them to stop in perfectly good Eskimo that I had learned; but it was no use.

The dogs ran like the wind. I ran too for a short way, marvelling that I could keep my feet when my toes hurt me as they did. Then I went down. An ill-advised stumble, a slippery spot and a slide, and I was done. But I held to the traces. I could not resign myself to seven days of waiting while someone else caught those dogs again.

They dragged me practically across Shugliak's nineteen-thousand-square-mile expanse during the next ten minutes. We, or at least I, went through nearly everything life in the Arctic has to offer, I should say, losing bits of my very being as I sped along, but not, you must be so kind as to remember, losing my dogs.

As we went through pools of water I was perfectly conscious of being responsible for the fans of spray that spread out at either side. I made special sounds whenever I passed over rough ice, and thumped and whacked noisily in getting across chasms. The dogs paid little attention to me. As a matter of serious fact, I am glad they didn't. Had they turned upon me, smelled or tasted the blood that oozed from my wrists and chin, and noted that I was all but done in, they might just possibly have made an end of me. There are deaths in afternoons that Ernest Hemingway knows nothing about.

We finally got back to the *komatik*. Jack said, "You're bleeding, Doc!" and wiped me off with his handkerchief. And then we all laughed. Odd that laughter should break out at such preposterous moments; but it does.

I was peacefully drowsy during the rest of the journey. When we reached the opposite side of the inlet we realized the dogs could never pull the *komatik* across the gravel and moss and mud of the snowless tundra.

We untied the tent and made camp. The place was alive with birds. I knew not which direction to turn, there was such a bewildering clamor from jaegers, old-squaw ducks, loons, and geese. A pair of small geese had a nest on an island in a little pond near us. We spied a crane walking solemnly about in the distance. Swans were flying overhead.

KOOGZHOOK: WHISTLING SWANS

I was much too excited to make tea. That afternoon I saw my first occupied nest of a whistling swan, a great mass of moss, roots, and grass, three feet high and seven or eight feet across at the base, and lined warmly with snowy down. In the nest were three huge, long eggs.

These eggs I collected. When I got back to camp I found that my companions were afield. I was ravenous. I got out my little black box of oölogical instruments, blew out my three precious specimens and dined, in solitary bliss, on four tin cupfuls of raw swan's egg.

·   ·   ·   ·   ·

The story of our several days of bird-collecting at Itiua-chuk would be a long story. I was so busy I slept but little.

We decided to return to the post by way of Prairie Point. This would give us opportunity to see more birds, and we would avoid the blueness of river-mouths.

At Prairie Point I did my best to catch up on sleep, but had little success. There was too much to be done. A snow bunting persisted in singing from the ridge of my tent. I could not bear to kill the little thing, but his song became a veritable screaming in my ears. Finally I took to snapping the bottoms of his feet (that I could see through the canvas along the ridge-pole) with a little rubber band, and this persuaded him to perform somewhere else.

Most lovely of the birds at Prairie Point was *Ahigeriatsuk*,[1] the rosy-breasted Sabine's Gull. A colony of these gulls nested not far from my tent.

We *komatikked* back to the post before the ice of the inlet had rotted badly.

---

[1] *Xema sabini.*

CHAPTER XLVI

Birds, Butterflies, Mosquitoes and Swimming Lessons

WE HAD two or three days of Spring, days like the one I have just told you about.

The Arctic Spring is an interesting season, but it wears you out. All this unceasing noise, all this breaking through ice, all this trying to walk on frozen lake-bottom, all this fording of swollen streams—honestly, if there were more than two or three days of Arctic Spring you would not live to see the Arctic Summer.

I was so busy with these singing, courting, migrating, and nesting birds that I had a well-defined case of a rare mental disorder: *dementia tundra*. I never knew what hour of the day it was. How could I, skinning and writing and painting away without lamplight at one o'clock in the morning? I never knew what day of the week it was. Sam or Jack or somebody else always had to tell me. I would "breakfast" at six o'clock in the evening and get back from my "morning's" collecting trip at ten P.M. Sometimes, miles from the post, I would hear the Mission bell ringing and would know it was meal-time—but which meal? I would lie down for naps on the moss, trying to keep myself in some sort of "good trim." Sam gave up expecting me at any time. There was no such thing as schedule for "the Doctor."

One day, after a patient stalk, I shot a snowy owl. As I picked the specimen up I found myself suddenly very hun-

gry. I decided to go straight back to Sam and get something
to eat. I had my collecting creel,[1] gun, and several specimens.
Eagerly I began fording a stream. By the time I was three-
quarters of the way across I perceived that the real crossing
was yet ahead of me. By this time I was so impatient that I
could not see myself turning back. Another step and I was in
swift water up to my armpits. Still I did not turn back. I
swam. Everything about me was soaking wet in a moment,
and the owl was a sorry sight by the time I pulled up on the
bank.

Did you ever try swimming with a gun and a snowy owl
and a collecting creel? You hold the owl's wing in your teeth
in the manner of a Chesapeake retriever; the gun in your
left hand more or less up out of the water; and the creel
takes care of itself. You effect a side stroke, kick considerably,
and move forward somehow.

I became interested in the pretty Lepidoptera that flitted
up from the moss and grass: dull yellow sulphurs; brown
fritillaries; tiny glaucous-gray azures that were exceedingly
hard to see; and various small moths. I gave whole days over
to butterfly chasing. Sheltered slopes to the south of the
higher ridges were the best collecting-grounds. Here the insects flew about whenever the sun was shining. But the moment the sun went under a cloud *not a butterfly was to be
seen anywhere.*

I kept seeing a sort of butterfly I had not collected, a rapid
flier with checkered wings something like the species I had,
as a boy, called the Lord Baltimore. In trying to catch one of
these elusive creatures I sprained or broke my right ankle,
and the gay Allegretto tune I had been singing changed all
at once to a Largo. My case of *dementia tundra* entered another phase. I crawled where I had leaped and run.

[1] A fish-creel is an excellent collecting-basket.

Philosophy, that saddest of all necessities, came to my aid. I gathered flowers and weeds instead of butterflies and birds, hobbling from plant to plant, garnering in veritable haycocks of weeds and sedges and bog-cottons. My workroom filled with boulders weighing down plant presses.

I found a certain sort of dock with arrow-shaped leaves and red flower-stalks that had a pleasant, sourish flavor. Many of these dock-plants I collected and preserved; but most of them I consumed. Santiana showed me also the edible roots of a small, leguminous plant.

We were not much bothered by mosquitoes near the coast. The season was rather drier than usual, and there was so much cool wind from the ice-filled bay that the "flies" never really swarmed about the post. In the sheltered marshes of the interior the prolific creatures rose in clouds, however, making life miserable.

Let us dismiss the subject of mosquitoes as promptly as possible. As mosquitoes are the curse of the North Country's summer, so they are the curse of North Country memories. You recall frozen nose and sprained ankle and arsenic poisoning and snow-blindness with a degree of amusement; but you recall the "flies" with a groan and an oath. Midsummer, the so-called "fly-season" of the Arctic, would be one of the loveliest seasons in the world were it not for the "flies." Why mention them further? What can be said or thought that will do any good anyway?

. . . . .

In latter June, Tommy Bluce and his family left for the Cape Kendall region. Tommy took one of my shot-guns, for he was to get some blue goose specimens for me. It would have given me great pleasure to go with Tommy to this famous *khavik* breeding-ground, but I knew that a thorough collection of birds could best be made by my staying at one place and working intensively. I was delighted at finding a few blue geese nesting about Coral Inlet. Some of these had a peculiar piebald coloration, white splotching here and there that indicated possible hybridization with the structurally similar lesser snow goose, a species that nested also near the post.

KHANGHUK: BLUE GEESE

Tommy returned early in July, bringing with him several fine skins and many eggs from one of the two known nesting-grounds of the blue goose.[1] He had had a hard trip, fording the swollen streams. In establishing camp near the *khavik-*colony he had had no end of trouble finding a place dry enough for the *tupek.* On this dry spot the marooned lemmings gathered in such hordes that the children spent all their time killing them.

Tommy had skinned the geese neatly. He had stuffed them with grass and moss; he had sewed them up carefully; he even had crossed their legs and attached neat paper labels in the most up-to-date museum style. Alas! he had put no pre-

[1] The other nesting-ground of the blue goose, discovered in 1929, is in south-western Baffin Island.

servative whatsoever upon the skins themselves. Already there was an odor of mould and decay. I had to take all the specimens apart and do the best I could at salting them "down." The eggs had been blown in the field with a hollow grass. I was disappointed that these *khavik* specimens were not perfect. But breeding specimens of blue geese were rare specimens, and the eggs all but unknown, so I could not complain.

. . . . .

On July 15, the ice blew out of South Bay. Summer had won at last. Motor-boats were hauled back to the water. The floating-wharf was replaced. Everybody talked of trips to Walrus Island, to Koodlootook River, and to Cape Low.

. . . . .

Muckik told us that he and his family were going by motor-boat to Chesterfield Inlet, on the mainland, to get a wife for Kooshooak. Kooshooak, you remember, was Muckik's son. Seventy miles across the open sea of Sir Thomas Roe's Welcome!

. . . . .

I was so busy being scientist these days that I wasn't much of a human being. I turned grind in spite of my high resolves. I had dozens of specimens of all sorts to prepare every day, photographs to take, and water-color sketches to make. It would not be long, now, before the *Nascopie* should come to take me away. Sam said he had had a radio message saying that the boat had left Montreal on July 13.

All was not drudgery, however. When my ankle became usable, I remembered the bathing-suit Jack had given me and decided to go for a swim. And Jack joined me. Attired in our red- and orange-colored handkerchiefs we ran down to the floating-wharf and dived in.

Have you ever plunged into really icy salt water? I mean salt water that has ice-chunks actually floating round in it? Do you remember the way you gasp and gag and the way your skin burns?

We dived in six or eight times, not swimming much, and then ran back to the house. It was invigorating exercise; but it was also enervating. Our skin glowed after the rub-down, but our eyes went heavy. There was nothing to do but take a nap.

The Eskimos became vastly interested in these natatorial interludes. Word would go round, in mid-afternoon, that "Jack and the Doctor" were going in swimming, and every Eskimo in the place would line up on the shore waiting for those pinky bodies to burst from the door, run down to the wharf, and disappear into the green sea. They cheered lustily when we came up. They knew we were having a grand time showing off. And they were not going to let applause be anything but hilarious.

Some of the men said they wanted to learn to swim. Kayak-juak apparently could swim a little already and he wanted to dive. One day several of us decided to go in together. We paddled in canoes out to one of the motor-boats. I tried some fancy stunts: a back dive or two, a swan, and a near jack-knife. Kayakjuak tried diving. He splashed a good deal but got on fairly well. One of the other men appeared in an odd drapery of inflated seal-intestines: water wings of a sort that I daresay never had been tried before. Santiana merely let himself slowly into the water, hanging onto the boat and looking completely miserable.

CHAPTER XLVII

Ooyalak the Aivilik: the Story of an Eskimo Baby

In TELLING the story of Ooyalak the Aivilik I must go back
a way: back to the Easter festival in April.

It was at this festival that Jack and I first saw Ooyalak,
Pumyook's son, a baby three months old. Snug in the damp
warmth of his mother's *kooletah*-hood, or pressed close to
her breast as she lay under thick caribou-skins, Ooyalak had
been brought by *komatik* across the barrens and frozen bay
from Native Point. And because his own mother and father
were a little poor he had been adopted into the "rich" family of John Ell, at the post. Here Mary Ell, his new mother,
carried him about in her *kooletah*-hood. For Peter, Mary
Ell's other young son, a lad about whom I have told you,
was now able to run about by himself.

When Jack and I first saw Ooyalak, there at the Easter
festival, we did not call him beautiful. We contemplated him
much as we had contemplated embryonic lemmings and the
bodies of skinned Arctic foxes. He was so young that he did
not often open his eyes. Most of his time, according to our
observations, was spent in slobbering, in squirming, in threatening to cry, or, to use Jack's words, in "doing other unusual
things." Both of us learned by degrees that a yawn or a certain kind of squirming often portended disaster; and at such
ominous moments we were more than eager to concede to
Mary Ell the right to care for her charge in any way she
pleased.

We touched, examined, or carried Ooyalak at every opportunity. When, after the passing of a month, he had learned
to coo and smile at our noisy inanities, we began to love him.
Eager for entertainment in the loneliness of our shut-in evenings, we begged Mary Ell to let us carry him about or play
with him on the bed; and Mary Ell, with a shy movement of
her head, and a peculiar squeal of surprise at our interest,
usually gave in to our importunity and lifted the baby from
his deep retreat.

As a rule he was naked, save for a comical hood of dingy
flowered silk, edged with limp lace, that framed his round
face neatly. He was obviously not a full-blooded Eskimo, for
his skin was not dark; but his hair was so black and his large
eyes of such lustrous brown that we instinctively talked of
the day when he should become a great hunter of *Aiviuk* and
Nanook. His legs and arms were ever on the move and his
hands were strong. On his back, especially along the lower
part of the spinal column, were dull bluish-gray blotches that
the Eskimos gazed upon with great satisfaction. Pointing to
these areas of strange pigmentation they were wont to ex-
claim: "He is fair-skinned, but nevertheless he is of the Innuit. He will be *angoti-marik* [1] when he is full-grown."

Already, it was explained to us, Ooyalak's future wife had
been selected. She was the baby daughter of a hunter living
in the Repulse Bay country.

As our interest in and love for the child grew we found we
could no longer call him Ooyalak. We began calling him
Albert. Why we called him Albert I cannot say; but the giving of this new name was important to us. As Ooyalak he
seemed to belong to that part of the Eskimo world whose
mysteries we could not penetrate; as Albert he was somehow
ours, and ours alone.

---

[1] A real man.

Albert developed an early aversion to housework of any sort. Perhaps he did not like being jostled about in the *kooletah*-hood as his mother clattered knives and plates in a dishpan. Perhaps a vague distaste for too much cleanliness disquieted him. Perhaps he merely learned that certain loud noises he, himself, was capable of making often produced pleasurable results. At any rate, about sixty seconds after Mary Ell had begun her domestic duties of the evening, he was wont to raise his voice in a lusty howl—a signal, we took it, for Jack and me to drop office work, bird-painting, lemming-skinning, or diary-writing, and rush instanter to the assistance of our young friend. Mary Ell must have sensed that her child was becoming wilful; but what was she to do against so formidable a triumvirate?

Abroad at last, obviously delighted at his escape from the hood, and eager to experience our gentle maulings, Albert nevertheless felt it his duty to cry awhile longer, to prove, let us suppose, that his grief was of a deep-rooted and genuine sort, not to be soothed by the sudden appearance of two noisy anthropoids. Tears filled his great eyes, flooded his face and neck, and dampened his fists; but the flash of a mirror, a gruff word or two, or a little nuzzling in the neck set him to cooing, squealing and kicking ecstatically.

"*Now* how is our little tootilikookook?" [1] we would ask, so loudly that Mary Ell could hardly miss the note of exultation in our voices; and Albert, fist in mouth, would look at us wide-eyed for a fleeting instant, then turn his tear-washed face away as a smile spread from his eyes to his mouth and finally even to his ears—a smile so indescribably beautiful that Jack and I couldn't help looking at each other, solemn-

---

[1] Not an Innuit word; a word of the author's coining that became more or less popular among the Aiviliks as a synonym for *nootarak, baby.*

eyed, seeking relief from the strange sensations that ran up and down our spines.

"Look at him!" Jack would say. "He's built just like us, Doc, only he's so small." Obvious, of course; but my own thought processes somehow didn't carry me much farther.

"Wouldn't it be great if he were really ours?" I rejoined, feeling myself becoming lonely and somewhat silly.

. . . . .

The June sun struck at the heart of the snow-drifts. Water dripped from the roof of the servants' house. *Amauligak*, the bunting, returned from the South in a whirl of soft snow, pitiably thin and eager to find shelter near us, even among the ever-hungry dogs. You remember how the lakes thawed, and how the purple and yellow and white flowers opened to the all but ever-present sun. All this time Baby Albert was growing; not rapidly, as did the young foxes and caribou calves, but rapidly enough, we thought, to change his appearance ever so slightly for the better from day to day.

By the middle of July he had received a full set of clothing—not Eskimo clothing, but White Man's garb; a gingham dress, a funny coat, woollen socks, and tiny moccasins of cloth. So civilized was he that he did not even wear an amulet about his neck; and he continued to sport that gaily colored lace-edged hood.

During early August Jack and I were so busy preparing for the call of the *Nascopie* that we could not spend much time with our Eskimo friends. The packing of extensive collections and the completing of official reports did not keep us, however, from little daily interviews with Albert. He now was sucking candies and bones, gnawing at harpoon-ends and bedposts, sipping noisily at his toes, and balancing himself uncertainly on his bowed legs while clasping our thumbs with his tiny hands. What noises he made! Could he have expressed intelligibly the tumultuous thoughts that must have stormed his small brain, Mary Ell would probably have blamed us for making a heretic of her son.

One day we caught a young rock ptarmigan. The gentle bird crouched on the floor or walked about craning its neck toward the windows. It called in a mellow, almost dovelike voice. Its softness, its youngness, and the fresh clearness of its eyes reminded us of Albert, and we decided that the two youths should become friends.

When Albert saw the ptarmigan his big eyes sparkled, his arms waved boisterously, and his feet kicked more energetically than was their wont. Noting how eager he was to become acquainted with this new comrade, we lifted the bird to the level of his eyes and told him, in the usual baby-talk, to "see the little chickie."

Did Albert stroke the soft, handsomely patterned feathers of the meek head? Did he point to the bright eyes or the red combs above them and coo his approbation? Did he make the slightest gesture of affection? He did not. Shouting some new monosyllable which must have meant something like, "Down with all Ptarmigan!" he lifted his clenched fists, kicked his legs anew most savagely, and dealt the innocent grouse such a stunning blow that a dozen small feathers drifted outward and down to the floor.

"Mustn't treat the bird so rough," we warned, forgetting our grammar for the nonce, actually fearful that the bird was done for.

"Glah-ah-oh," answered Albert in a shout, his brown eyes flaming.

And we put the dizzy ptarmigan in the adjoining room.

·   ·   ·   ·   ·

In the second week of August, Muckik's family returned from Chesterfield. We were glad to welcome our old friends home. We were glad to see Kooshooak and his new, squint-eyed mate walking hand in hand along the gravel-beach. But Muckik's return was, alas, a tragic return. For he brought with him the dread germs of an influenza that had been raging through the Eskimo ranks on the mainland. Muckik himself had contracted the disease. His wife had almost died of it. His family were so thin and pale they were scarcely recognizable. They were coughing hideously.

Muckik's family were so weak we felt we could not send them away; even though we realized that it was "ship-time" and that all the Eskimos of the island were on their way, at that very moment, to the post.

Knowing that a malady of this sort will spread like wildfire through Arctic communities where no resistance has been developed, we tried to explain the habits of germs, gave advice as to cleanliness, warned of the mortal danger, went through our small supply of medicines, and, realizing full well that our words were not likely to be understood and even less likely to be heeded in view of the Eskimo's traditional sociability, awaited the worst.

Aivilikmiut and Okomiut, all the Innuit of the Island, came to the post. The *tupek*-village sprang up along the shore. We did our best to keep Muckik's family apart. But we failed.

Soon the ablest men fell ill. Sniffling, coughing, and choking were to be heard on every hand. Jasper's wife almost died. Cabin Boy became so weak he could not rise. Shookalook dragged himself about, his mighty frame racked by deep, almost incessant barking. For two nights I lay in a tent pitched in the midst of the Eskimo encampment; but sleep I could not—there was too much noise. Coughing, moaning and groaning combined into a ghastly chorus. I did not want to sleep for I feared that the children might suddenly go into convulsions.

Baby Albert contracted the malady within three days. He cried pitiably; his fever ran high; his eyes became wretchedly sore. It was painful to look upon his gray, drawn face. We did our best to help him. We washed his eyes, did all we could to see that he was kept clean and dry, and made certain that Mary Ell was caring for him as well as she could.

Almost frantically we awaited the supply-ship with its store of fresh food, its ample stock of medicines, and, most of all, its experienced doctor. Every evening we listened to the radio for some word of assurance that help would come quickly. None of us white men became sick.

On Friday, August 15, the *Nascopie* arrived. Our little community bustled with nervous excitement. In the bright, healing weather some of the Eskimos had recovered from their illness. The doctor diagnosed the epidemic as influenza in a mild form. But he told us that Baby Albert would die.

Jack and I accepted the doctor's words in silence. We had seen so much that we could not simulate any surprise nor give voice to any feeble hope. We went about our work in a strange, uncommunicative mood. How queer that we should be well; that we should go ahead drinking water and eating food at meal-time; that we should waken in the morning conscious of having slept soundly! The doctor had assured us there was nothing, absolutely nothing, we could do for the baby. Drugs could not be administered for Albert was too weak. We went ahead with our packing.

Those who knew us best probably had no idea of the unfamiliar sorrow that made us quiet for a time, then suddenly boisterous; that made us talk gaily of the Great Outside to which I was returning, then lapse into sudden, almost hostile, silence. There were moments when I hated Jack for the very calmness of his forehead, for the brightness of his eyes. I shrank from revealing my feelings to him. I despised myself for being so ignorant and impotent. Surely I was as intelligent as the doctor! Surely there must be something we could do, for the baby was still alive. We went ahead working, thinking of heroic measures, pondering upon miracles—but doing not a thing for Baby Albert. And the hours stalked by.

That evening I saw Jack standing at the corner of the house. He was looking toward the floating wharf. I, too, looked toward the wharf and ascertained that there was nothing there worth more than a second's contemplation. So I walked over to him, looked in the same direction he was looking, stuck my hands in my pockets, and stood there.

"I wonder how little Albert is."

I'm not sure that Jack made any response. But we both turned and walked toward the servants' house. The gravel crunched loudly underfoot. We turned the greasy knob of the outer door, pushed open the heavy, squeaking, inner door, and made our way across the dim, hot, ill-smelling kitchen-living-room to the small bedroom adjoining.

There, on the floor, on a crude pile of blankets, lay Baby Albert, Mary Ell huddled at his side. His face was scarcely that of the living. His eyes, partly closed, were hidden in purple shadows cast by the gray brows; their rich lustre was gone; their large pupils were directed at nothing. His breathing was hard and irregular. At times he squirmed and cried out convulsively, trying to get air. There was no doubt that he was dying. We asked our friends whether we could get them anything or help them in any way.

"*Naga*," [1] they said quietly.

As we went out, brown, trusting faces smiled upon us

---

[1] "No, there is nothing that can be done."

quietly. They knew that we were to leave them on the morrow, perhaps never to return. Some of them knew also that the baby had not long to live.

We went at our unimportant labors, glad to be out in the sunshine and air, but oppressed by the feeling that we, too, should be at the side of the pile of blankets watching the sick child to the last. It seemed scarcely fair to our friends that we should move about so briskly, preparing for our departure.

. . . . .

On the following morning Jack and I looked across the harbor to the distant cliff at Itiujuak and the purple barrens beyond Itiuachuk, realizing that this was to be our last day on Shugliak. We ate our breakfast of toast and coffee, not in awkward silence, but talking quietly about homely matters, recalling the good times we had had together through the winter. We had learned to be honest with each other. "I'm afraid Albert is going to die today," I said. "He was very weak yesterday. He could hardly breathe." When Jack answered "Yes," I knew that we must go to the servants' house at once.

When we entered the familiar, ugly room, there was a reassuring bustle at the table where One-eyed Joe's wife was mixing bread. Two men, seated on the floor near a window, were eating *netchek*-meat from an old pan. Small children moved about quietly. The door to the little room at the side was closed. Near the door, her wrinkled, tattooed, but well-shaped hands held up to the sore at her neck, stood old Shoo Fly—ever busy, ever kindly, ever wise Shoo Fly, mother of our little island-community.

"Tell us how the baby is," we said.

Shoo Fly's answer was direct. Her hoarse, croaking voice

bore no trace of sorrow, worry, or panic, but her face looked older than usual. "Something must be wrong," she said. "He isn't breathing any more."

We opened the door to the bedroom slowly. The dead infant's real mother, Pumyook's wife, sat on her crossed legs on a trunk, her face directed toward the floor. At the side of the pile of blankets still on the floor crouched Mary Ell, the foster mother, her braided hair hiding her down-turned face. On the edge of the iron bed sat Pumyook, a gray-brown handkerchief wadded into his right hand. Back of him lay three girls. The baby was covered with a soiled, heavy blanket.

I was amazed that the room was so very quiet. About Jack and me seemed to be not living beings, but statues. Our friends did not even glance at us. Then Mary Ell's shoulders shook abruptly, and a quavering, thin, high wail rose from her bowed head. So restrained yet so penetrating was this long-drawn-out moan of despair that it seemed anything but human. It reminded me of the cry of *Ookalik*, the hare, caught in the viselike talons of an owl; or of the distant moan of a starving dog trying to pull itself from the jaws of a trap into which it has blundered. Jack must have felt about as I did. We wanted to sympathize with the sorrowing woman; but the sound of her crying seemed to declare us creatures of another species, incapable of communicating our feelings to one whose innermost soul we simply had never known.

Jack held a handkerchief near the bowed head. A brown hand grasped the bit of cloth, and we could tell that Mary Ell was wiping her eyes; but she did not look at us. Shoo Fly came in, gave us certain details concerning the illness and death of the baby and stood waiting as if she expected us to leave. Abruptly she went out and One-eyed Joe's wife came in, seated herself on the trunk near the baby's actual mother, and spoke several sentences in a friendly voice.

My Eskimo vocabulary was so limited that I asked Jack to tell our friends that we were sorry, and that we wanted to do something to show our friendliness. One-eyed Joe's wife explained that the baby had been very sick just before it died and that it was lying in a dirty place.

I felt exceedingly shy in the presence of my friends, the more so because I could not say much to them; but I asked Jack whether he would mind helping me move the body to a cleaner spot. He spread a white handkerchief on a dry portion of the blanket-bed and I lifted the thin corpse to a new resting place. As I adjusted the small head so as to give it the appearance of more perfect ease, and covered the gray face with a corner of the blanket, Mary Ell broke out into renewed wailing, and I was stricken with terror lest we had somehow offended. But Jack and I felt better for having made one physical gesture in an attempt to help our friends. As I rose from the side of the corpse a dog outside the window began to howl. Other dogs joined until the weird cry was taken up by every animal at the post. Jack looked at me oddly; and I marvelled that at such a moment the dog chorus, which was usually inspired only by the Mission bell's ringing or by a fight among the *komatik*-team leaders, should begin now. I was not comfortable.

The door opened and old Munnapik came in. Munnapik was related to both the actual and the foster mother, and to the father. His shining face was not more than usually stern. Standing near the dead baby he addressed us all in a clear, matter-of-fact voice. "There is no reason for being sad. You women must not cry so much. See! you are making the dogs howl. We must not be sorrowful over the death of a baby. It is much better for a baby to die than for a grown person to

die. There are always many babies. Pumyook's wife, even
now, has another baby inside her. There are still many grown
men among us and our dogs are well."

John Ell grasped Mary Ell's arm and lifted her from her
place at the side of the blanket-bed. Choking, she left the
room. "I didn't want to take the baby in the first place," she
said. "Now they will blame me for its death."

I wondered, as I stood there, whether Jack and I were not
the saddest persons in the dark room. Had Mary Ell been
crying because she missed the baby, or was it because she felt
she would be blamed for the death of a hunter-to-be? Had
Pumyook's wife been sad at all? She had not wept much nor
loudly. Had the father been actually sad, or were his nerves
merely overstrained? Once again I felt that my Eskimo
friends and I belonged to different worlds. I felt incapable
of sympathizing because I was incapable of understanding.

Within a few hours Jack and I were to leave on the *Nascopie*. Among the noisy throng on the beach would gather all of
the dead infant's relatives, some of them talking and laughing gaily, all of them smiling a farewell.

They would bury Baby Albert on the morrow. They would
carry the body to the cemetery beyond the Mission and cover
it crudely with a pile of stones. Perhaps they would lay some
trinkets beside it in accordance with an ancient custom. No
flowers would be put on the grave. The Mission Fathers
might erect a thin, black, wooden cross. Only the hungry
dogs would keep a night-time vigil. And only the weight of
the stones would keep these dogs from devouring the corpse
—skin, flesh, and bone.

But *Amauligak*, the snow bunting, would mount the topmost pinnacle of the grave and sing loudly his late summer
song, the feathers of his white throat pulsing and quivering;
and *Teggeuk*, the silken-coated weasel, would slip noiselessly

through the stones, all over and about the small carcass, trying to fathom the mystery of every unfamiliar and fascinating odor, touching the dead nose and the dead fingers and even the dirty, flowered-silk hood with the tips of his dainty whiskers.

CHAPTER XLVIII

The *Nascopie*

I SUPPOSE I was glad to see the *Nascopie* on that morning
of August 15, 1930. Assuredly our first glimpse of her smoke
—a little smear of gray on the horizon—gave me a sensation
I had never before experienced. It was good to know there
was a doctor aboard who would help the sick Eskimos. It was
good to think of the mail from home; of the dainty foods
(butter, potatoes, perhaps even some apples or celery!); of
the new books and magazines. It would be jolly to see Captain Murray again, and all the H. B. C. officials.

But when the *Nascopie's* engines stopped and the great
anchors went down; when the White Man's customary greet-
ings had been said all the way round; when, confronted at
last with complete realization that "the ship" was here, I
knew not what to say nor with whom to talk—a pang stabbed
me through. All at once I wanted not to see any one. All at
once I wanted to go straight to my workroom and sit on the
familiar green blanket of my cot. Old Shugliak! A little
while and I should be leaving this island that had somehow
become my home! These happy beings who had been my
companions day after day after day; these Eskimos; these
gruff dogs—I had come to love them all. This unquestioning
friendliness of the gentleman named Sam Ford! This brave
wilderness of rock and ice and sea and sky that had been my
workshop and playground for a year! Yes, it would be hard

to leave. It would be hard to adjust myself to inane correspondence that leads nowhere; to telephone bells that wreck conversation and kill Art; to the committee meetings, conventions, and civilized confusions to which I must return. I should have to be brave like this island. None of this keep-your-chin-up stuff; none of this stiff-upper-lip stuff; something more basic than that. And I should have to remember that there are times when it is best to *not-think.*

Jack's mother and little brother and sister were on the *Nascopie.* They had come to spend the winter. And Jack was to leave, on this very *Nascopie* that had brought his loved ones to him. A few hours together: trifling conversation about this and that wedged in between official talk; a kiss or two; a meal together near the big, square stove; a "My, how you've grown, Jack!"; an "Honest, you're prettier than I ever seen you, Mother!"; a spasm of loneliness not revealed in tears or solemn words or breaking voice, but a spasm of loneliness nevertheless—and "Good-bye!"

Gladly I recalled that a little packing yet was to be done. I grabbed hammer and nails and an armful of excelsior and ran across the gravel to the store. The antlers of the big bull caribou wouldn't go into the box without the most diplomatic coaxing. There was that mess of a loon nest I had been drying out, those delicate mushroom specimens in pasteboard boxes. "Don't forget the drum Khagak gives you, Doc!" Jack said.

A tall young Englishman by the name of Jimmie Drummond-Hay, who was among the *Nascopie's* passengers, offered to help me. We had a pleasant time chatting about the island.

At last came the hour of departure. The Eskimos all said *"Tugvahootit"* to me, shaking my hand in their funny, brusque way, all of them a trifle embarrassed, all of them wondering whether to be solemn or gay at "ship-time," all of them smiling.

It was "*Tugvahootit*, Doctor!" "*Tugvahootit!* Be good, John Ell!"

"*Tugvahootit*, Muckik!"

"*Tugvahootit*, Tommy Bluce!"

"*Tugvahootit*, Kayakjuak!"

"*Tugvahootit*, Scotch Tom and good old Shoo Fly!"

"*Tugvahootit*, Kooshooak, you young devil!"

"*Tugvahootit*, Mary Ell! You certainly took good care of us, Mary Ell!"

"*Tugvahootit*, pretty, black-eyed Ookpik! Are you still a little shy when in front of the gray-eyed White Man?"

"*Tugvahootit*, Khagak, with your mygodboy accordion under your arm!"

"*Tugvahootit*, Santiana!"

"*Tugvahootit*, Little Pete and squint-eyed Mikkitoo and sad-faced Tooghak!"

"*Tugvahootit*, Widow Kuklik!"

"*Tugvahootit*, Mrs. One-eyed Joe!" . . .

Yes, it was *Tugvahootit* at last to Shugliak; *Tugvahootit* to The Island-Pup That Is Suckling the Continent Mother-Dog!

But it was "*Au revoir!*" to the Mission Fathers, and a hearty "*Merci!*" for courtesies extended during the winter.

And it was "Good-bye, Sam! You've given me a grand time, Sam. Hope the winter goes well with you and the family, and with the people at the servants' house and with all the others. Let me hear from you, sure now. I'll talk to you over the radio when I get back to the States. Honest, Sam, you've given me a year I'll never forget. We got some good work done too, didn't we? You know I couldn't have done a thing or even turned around on this island without you, Sam!"

. . . . .

Jack and I stood on the deck of the *Nascopie* as she weighed anchor and steamed slowly out of the Inlet. There was the post with the big Company sign and the spire of the Mission building. There was the kitchen window of Sam's house, and the water-barrel, and the crack where the buntings had their nest. There was the little powder magazine and the oil-shed. There were the thin crosses of the cemetery where I had seen the weasel and where Baby Albert would be buried. There, a little to the west of us, was Munnimunnek. The red roofs of the post were very small, now. We couldn't even see the crosses of the cemetery. To our right rose gray *noovoodlik*, the "peculiar shedlike hill" of the charts.

"Look, Doc! There's the place these Baffin Landers got their three bears!"

"Look, Jack! There's a *netchek*, almost under us. Let's whistle and see what he does!"

Jack and I were glad to be together. It was a little as if we weren't parting with the island all at once. Too, this *Nascopie* was rather a new world for both of us. New persons to meet. New foods to eat. New magazines to read. We could face this world together, for a while.

At dinner in the softly lighted saloon that night there was conversation about many matters. Somehow I found myself thinking hunting and fishing and dog-driving and butterfly-catching instead of Literature and the Arts. But Literature and the Arts were with us. I told my companions something about the ten books I had taken with me for winter reading. And Jack piped up (I swear I shall never forget Jack as long as I live): "Well, you remembers, Doc, what I likes best is the American Literature. I likes Walt Whitman!"

Jack and I stood watching the dim brown band of barren island to the north of us as long as we could see it. Darkness fell. Fog drifted in from Sir Thomas Roe's Welcome. And we remembered that the spirits were sad, hiding themselves behind the white vapors.